An Introduction to Time-of-Flight Secondary Ion Mass Spectrometry (ToF-SIMS) and its Application to Materials Science

An Introduction to Time-of-Flight Secondary Ion Mass Spectrometry (ToF-SIMS) and its Application to Materials Science

Sarah Fearn

Department of Materials, Imperial College, London SW7 2AZ

Morgan & Claypool Publishers

Rights & Permissions
To obtain permission to re-use copyrighted material from Morgan & Claypool Publishers, please contact info@morganclaypool.com.

ISBN   978-1-6817-4088-1 (ebook)
ISBN   978-1-6817-4024-9 (print)
ISBN   978-1-6817-4216-8 (mobi)

DOI   10.1088/978-1-6817-4088-1

Version: 20151001

IOP Concise Physics
ISSN 2053-2571 (online)
ISSN 2054-7307 (print)

A Morgan & Claypool publication as part of IOP Concise Physics
Published by Morgan & Claypool Publishers, 40 Oak Drive, San Rafael, CA, 94903, USA

IOP Publishing, Temple Circus, Temple Way, Bristol BS1 6HG, UK

This book is dedicated to M F and M C A Tsamados.

Contents

Preface   ix

Author biography   x

1   Introduction   **1-1**
1.1   Overview   1-2
1.2   Basic principles   1-3
      References   1-6

2   Practical requirements   **2-1**
2.1   Ion generation   2-1
2.2   Primary and sputter ion beam sources   2-2
2.3   Mass analysis   2-5
2.4   Ion detection   2-8
2.5   Ultra high vacuum   2-8
      References   2-10

3   Modes of analysis   **3-1**
3.1   High-resolution mass spectra   3-1
3.2   Depth profiling   3-3
      References   3-5

4   Ion beam–target interactions   **4-1**
4.1   Ion beam induced atomic mixing   4-2
4.2   Beam induced surface roughening and uneven etching   4-2
4.3   Beam induced segregation   4-3
4.4   Other beam induced effects   4-5
4.5   Depth profiling with cluster ion beams   4-5
      References   4-8

5   Application to materials science   **5-1**
5.1   Biomaterials and tissue studies   5-1
5.2   Glass corrosion   5-8
5.3   Ceramic oxides   5-12

5.4	Semiconductor analysis	5-15
5.5	Organic electronics	5-18
	References	5-22

| **6** | **Summary** | **6-1** |

Preface

The main focus of this short book 'An Introduction to Time-of-Flight Secondary Ion Mass Spectrometry' is the operation and application of a dual beam time-of-flight secondary ion mass spectrometer. It has been written with the aim to inform and help final year undergraduates, first year PhD students, and anyone who may be starting out using secondary ion mass spectrometry as part of their research. The book gives a very basic introduction to some of the fundamentals of secondary ion formation and collection, along with a description of the main components that make up a modern dual beam instrument. There is also a clear description of how the dual beam system operates to generate the various types of analysis possible: mass spectrometry, ion mapping, depth profiling, and 3D analyses. The final section of the book focuses on applications. Several different material groups are looked at, and as well as conventional ways of carrying out SIMS analyses, different approaches to sample preparation are also highlighted to emphasise the flexibility of SIMS as a materials characterisation technique.

Author biography

Sarah Fearn

Sarah Fearn obtained her PhD from the Department of Materials, Imperial College in 2000. After working in commercial SIMS analysis for two years, she returned to academia in 2002. Since then she has worked on a variety of research projects that have all applied SIMS as the main characterisation tool for understanding a range of material science issues such as glass corrosion, to calcium deposits in eye tissue. Over the last 15 years she has published ~50 papers spanning a wide range of material science issues and the application of SIMS. Since 2012 she has been in charge of the surface analysis laboratory in the Department of Materials at Imperial College.

An Introduction to Time-of-Flight Secondary Ion
Mass Spectrometry (ToF-SIMS) and its Application
to Materials Science

Sarah Fearn

Chapter 1

Introduction

For the last 50 years, secondary ion mass spectrometry (SIMS) has been at the forefront of high-resolution materials analysis and characterisation. A combination of factors makes SIMS unique amongst the analytical techniques widely available: it is able to measure all elements of the periodic table from H to U, along with isotopes and molecular species. Under optimum conditions the technique has extremely high sensitivity down to ppm (and in some cases ppb) coupled with very high surface specificity on the order of nm. This range of capabilities means that SIMS has been exploited in many wide-ranging areas of research and materials development. From its initial use as an instrument to investigate the material brought to Earth from the Moon missions of the 1960s, the application of SIMS can now be found in areas as diverse as conservation science to the development of renewable energy materials. In this introduction to time-of-flight secondary ion mass spectrometry (ToF-SIMS), an overall understanding of the main scientific principles underlying SIMS and its useful applications within the field of materials science will be given.

Over the years, SIMS instrumentation has dramatically changed since the earliest secondary ion mass spectrometers were first developed. Instruments were once dedicated to either the depth profiling of materials using high ion beam currents to analyse near surface to bulk regions of materials (dynamic SIMS), or time-of-flight instruments that produced complex mass spectra of the very outer-most surfaces of samples, using very low beam currents (static SIMS). Now, with the development of dual-beam instruments these two very distinct fields now overlap.

This book will highlight, in particular, the application of dual-beam ToF-SIMS for high-resolution surface analysis and characterisation of materials. Along with a brief overview of the underlying principles of the secondary ion mass spectrometry, there will also be some examples of how dual-beam ToF-SIMS is used to investigate a range of materials systems and properties.

1.1 Overview

Time-of-flight secondary ion mass spectrometry (ToF-SIMS) is a mass spectrometry technique used to analyse the chemistry of materials, *in vacuo*. An energetic beam of primary ions (0.1–20 keV) is used to bombard a sample surface. The bombarding primary ion produces a variety of sputtered particles: monatomic and polyatomic particles of the sample are produced along with electrons and photons and re-sputtered primary ions. The secondary ions that are formed carry negative, positive and neutral charges. The desired secondary ions are extracted and detected using mass spectrometry. A schematic of the instrumental components of a dual-beam ToF-SIMS are shown in figure 1.1. In a dual-beam system a primary ion gun is used to generate the secondary ions to be analysed under static ion beam conditions (explained in more detail in section 3.1) for high-resolution surface mass spectrometry. The second ion beam (known as the sputter gun) can be used for the controlled erosion of the sample, known as sputter depth profiling. The removal of material in this controlled manner enables the composition from the surface to the bulk to be analysed.

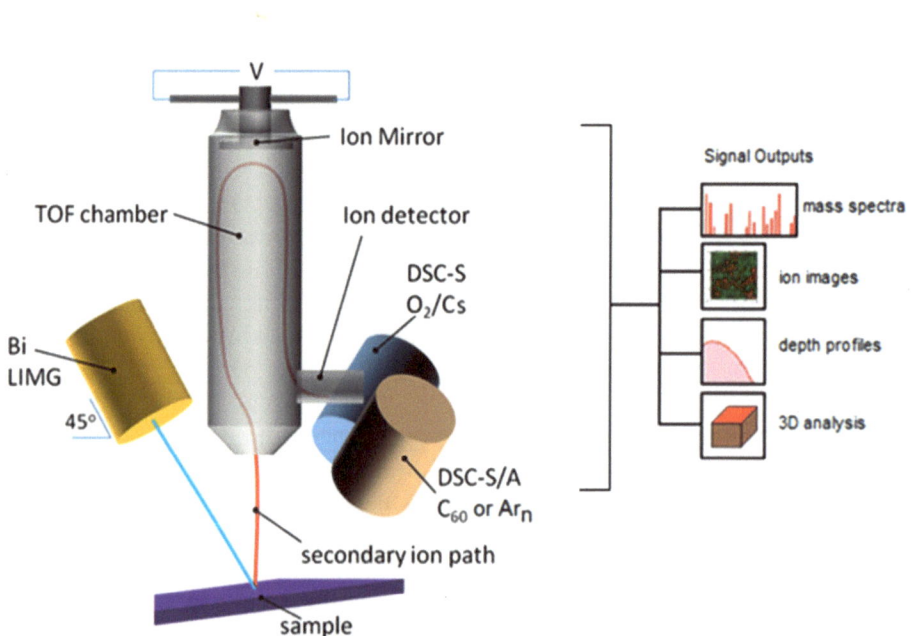

Figure 1.1. A schematic representation of the main components of a dual-beam time-of-flight secondary ion mass spectrometer. Secondary ions are sputtered from a target/sample by the primary ion gun. The sputtered secondary ions are extracted by an extraction potential into the flight tube and detected electronically, typically with a microchannel plate. Depending on the mode of operation of the instrument, a range of signal outputs can be obtained: mass spectra, ion images, depth profiles and 3D analyses.

1.2 Basic principles

Any atomic or molecular species that can be ionised and transported into a gas phase can be, in principle, analysed by mass spectrometry. At its basis secondary ion mass spectrometry, SIMS, is the measurement of the mass-to-charge ratio (m/z) of secondary ions generated from a target surface via ion beam bombardment. The formation of the secondary ions i.e. the ionisation, occurs at or very close to the emission of the particles from the surface. The ionisation process is, therefore, strongly influenced by the chemical state of the surface; this is known as the matrix effect. The ionisation and sputtering phenomena are complex processes, and a more in depth explanation of the mechanism can be found in the literature [1]. The emission of the secondary ions, however, can be described by the basic SIMS equation, shown below:

$$I_s^x = I_p C_x S \gamma F \tag{1.1}$$

where I_s^x is the secondary ion current of species x (i.e. the measured secondary ion counts of x), I_p is the primary ion beam current, C_x is the concentration of species x, S is the sputter ion yield of x and γ is the ionisation efficiency, i.e. the probability of the detected species forming positive or negative ions. Finally, F is the transmission of the analysis system.

The basic SIMS equation shows that the measured secondary ion counts, I_s^x, are directly proportional to the concentration, C_x, of the measured species, x, which implies that quantification can be easily made. However, as the measured secondary ion signal is also dependant on the chemical state of the surface where the ions have been emitted from, quantification is not so straightforward. The dependence of the ionisation of the measured species on the chemical state of the surface is, as previously stated, known as the matrix effect. This is a major hindrance to the quantification of many complex samples such as biological systems that have varying chemistry over the sample, and relevant standards are difficult to create. However, when standards can be accurately made, and the matrix is homogeneous as is the case for many silicon-based devices and semiconductors in general, quantification can be readily carried out with the use of implant standards. These are samples whereby a known ion implant of the element that is being measured is implanted to form a profile in the matrix material e.g. silicon.

The fundamental parameters of S, the sputter yield, and γ, the ionisation efficiency, are dependent on several factors relating to the selection of the ion beam and the properties of the target material. The sputter yield, S, is the amount of material removed during ion beam bombardment of the target. The sputtering yield is influenced, and increased, by the mass, charge and energy of the ion beam used. Put simply, a greater sputter yield will be obtained by a heavy bombarding ion beam species, as the energy is imparted closer to the sample surface. The properties of the target material, however, also influence the sputter yield and for any given ion beam energy the sputter yield for elements will vary by a factor of 3 to 5 over the periodic table, as shown by the early work by Wehner and co-workers [2]. An example of this

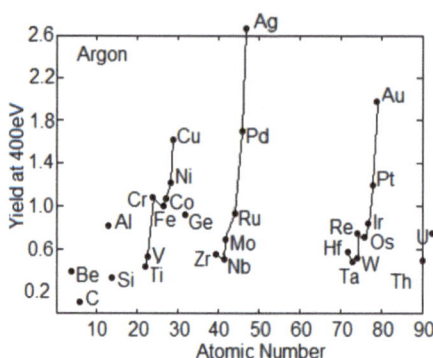

Figure 1.2. Sputtering yields for a 400 eV Ar ion energy for 28 elements versus the elements' atomic number. Reprinted from [2] with permission copyright 1961, AIP publishing LLC.

variation across the periodic table is shown in figure 1.2, whereby the yield variation due to Ar^+ ion beam sputtering was measured.

In covalent materials ion beam bombardment has a more damaging effect on the molecular and polymer structures and may rapidly destroy the chemical structure of interest. The amount of ion bombardment must therefore be limited to minimise the damage and another parameter is needed to monitor the sputtering. Along with the sputter ion yield, the disappearance cross-section, σ, is also monitored. The disappearance cross-section is an exponentially measured parameter and defined as the mean area damaged by one primary ion, and is related to the secondary ion intensity:

$$I_m = I_{mo} \exp\left(-\sigma I_p\right) \tag{1.2}$$

where I_m is the recorded signal of the molecular species of interest, I_{mo} is the original surface density of the respective molecular species, and I_p, the primary ion beam dose. Similarly to S, the disappearance cross-section also increases with the primary ion mass, energy and angle of incidence to the surface normal. The secondary ion efficiency, E, for a primary ion beam can thus be defined as:

$$E = \frac{S}{\sigma}. \tag{1.3}$$

For optimised primary ion beam conditions both E and S should, therefore, be optimised and σ minimised. Generally, it has been observed from spin coated monolayer polymer samples that for the commonly available ion beams, efficiencies scale: $Ga^+ < Au^+$ or $Bi^+ < Au_3^+$ or $Bi_3^+ < Bi_2^{3+} < C_{60}^+$ [3]. Brunelle also assessed the sputtering parameters using the Bi ion source on biological tissue and the values are summarised in table 1.1 [4].

Ion yields from organic materials increase as the primary ion projectile is changed from an atomic species such as Ga^+ to a cluster ion source such as C_{60}^+. This increase in ion yield is attributed to the way in which the cluster ion beam impacts upon the target surface. Upon impact, the cluster breaks up and the total energy of the ion

Table 1.1. Secondary ion yields, S, disappearance cross-section, σ, and ion bombardment efficiencies, E, for the bombardment of cholesterol in rat brain ([M-H]-; m/z 385).

Primary ion	Energy (keV)	S (10^{-4})	σ (10^{-13}) (cm^2)	E (10^8) (cm^2)
Bi_1^+	25	0.836	2.75	3.04
Bi_3^+	25	7.06	4.14	17.1
Bi_3^{2+}	50	9.91	3.52	2.81

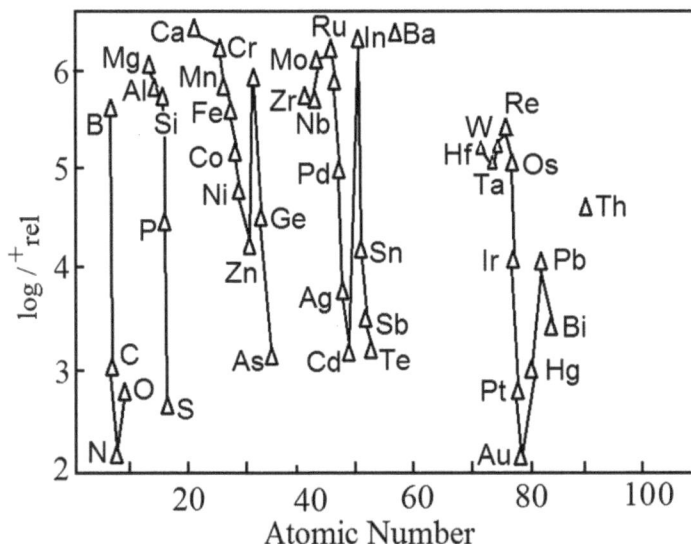

Figure 1.3. Variation of positive ion yield as a function of atomic number. Reprinted (adapted) from [5] with permission, copyright (1977) American Chemical Society.

beam is distributed over all of the atoms of the cluster, for example, a 20 KeV C_{60}^+ ion beam will result in an impact of only 666 eV per C atom. The penetration of the individual ion impacts results in less damage to the underlying structure of the sample, and a larger area of surface molecules are ejected from the surface. The erosion rate is also greatly increased so that any underlying damage is removed by the subsequent impacts. More recently the development of the Ar_n^+ ion cluster sources has dramatically improved the analyses of organic-based materials, and looks to surpass the use of C_{60}^+ ion beams. The development of ion beam sources and the effect on organic SIMS analyses will be discussed in more detail in section 2.2.

The ionisation efficiency, γ, is the ease or tendency of an element or molecule to form either a positive or negative secondary ion, and it is dependent on the ionisation energy or electron affinity respectively of the ion being measured. Elements in groups I and II of the periodic table have decreasing ionisation energies from the top to the bottom of the periodic table, and readily form positive ions: $X \Rightarrow X^+ + e$. Figure 1.3 highlights the variation of ionisation of the elements over the periodic table. Conversely, elements that have high electron affinities are more likely to form negative ions. Such elements are situated on the far right hand side of the periodic table, for example, the halogens readily form negative ions with fluorine being the most reactive.

References

[1] Sigmund P 1969 *Phys. Rev.* **184** 383
[2] Laegreid N and Wehner G K 1961 *J. Appl. Phys.* **32** 365
[3] Vickerman J C and Briggs D (ed) 2013 *ToF-SIMS: Materials Analysis by Mass Spectrometry* (Chichester: IM)
[4] Brunelle A, Touboul D and Laprevote O 2005 *J. Mass Spectrom.* **40** 985–99
[5] Storms H A, Brown K F and Stein J D 1977 *Anal. Chem.* **49** 2023

An Introduction to Time-of-Flight Secondary Ion
Mass Spectrometry (ToF-SIMS) and its Application
to Materials Science

Sarah Fearn

Chapter 2

Practical requirements

2.1 Ion generation

For secondary ion generation a number of processes need to occur. Firstly, the primary ion beam needs to be collimated and accelerated to an appropriately high energy. The impacting primary ion beam is then rastered over the target surface and causes elastic and inelastic collisions transferring some of the primary ion beam's energy to the particles in and around the surface (depicted by the grey lines shown within the sample in figure 2.1). During the ion beam bombardment of the target an atom or group of atoms may receive enough energy in a suitable direction enabling them to overcome the surface binding forces and be sputtered from the target, as shown by the blue atom in figure 2.1. These are the emitted secondary ions. This type of ion generation is known as desorption ionisation, and is the same phenomena that underpins the mass spectrometry techniques of matrix-assisted laser desorption ionisation (MALDI) and direct electron spray ionisation (DESI). The approximate depth of origin of the emitted atoms and ions during a SIMS analysis is often quoted as being around 2–3 atom layers [1] suggesting high surface sensitivity, however, due to the multiple collision processes that occur during sputtering this is often not the case. The location of the emitted particles can be up to 10 nm away from the initial impact of the primary ion [2]. The depth of emission of molecular species from organic samples may also vary, depending on the molecular species that is being monitored [3].

One of the main benefits of this ion generation process is that the sample can remain in its solid state, so that the chemical distribution within the sample is unchanged. The advantage of analysing samples in their native state means that along with the chemical information, the location of the chemistry is also obtained producing ion/chemical maps of the sample being analysed. This is in contrast to

Primary Ion

Secondary Ion

sample surface

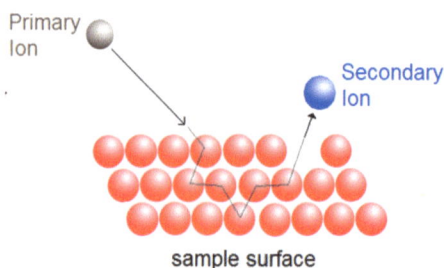

Figure 2.1. Schematic representation of a collision event that can lead to the formation of a liberated secondary ion capable of being captured in a detector. The primary ion (grey) impacts upon the sample surface (red) thereby transferring energy to the sample, which is then distributed through different atoms (grey lines). This process can lead to the ejection of a secondary ion from close to the surface (blue), which it is then possible to observe with an analyser.

other mass spectrometry techniques, whereby the samples must be changed into another form i.e. liquid or gas for the mass analysis to be carried out.

2.2 Primary and sputter ion beam sources

Over the last decade there have been significant developments in ion beam sources, most notably in the area of cluster ion beams. These new sources enable a much less destructive analysis of the softer/molecular materials, such as organic, polymeric and biological samples. The main primary ion beam on a dual-beam ToF-SIMS instrument is a liquid metal ion gun/source (LMIG/S). These ion sources produce a very focussed ion beam which translates into high lateral resolutions during secondary ion mapping. Originally, the LMIGs were based on gallium, then gold [4] but now the bismuth ion sources dominate [5] due to the improved secondary ion yields achieved with this source, in particular when used in the small cluster form of Bi_3^+ [6].

Along with LMIG primary ion sources, very large cluster sources based on carbon fullerenes have also shown potential for characterising bio-molecular structures [7, 8]. These electron impact (EI) sources work by using charged particles comprised of multiple atoms, commonly known as a 'cluster ion beam'. The main advantage of this type of source is that upon impact with the target surface, as the large cluster breaks up, the kinetic energy of the cluster ion beam is distributed over a large number of atoms [9–12] and multiple localised low energy impacts occur (i.e. the impact energy per ion is reduced) as opposed to one high energy impact from a single charged atom. The result of these multiple low energy impacts is that damage to the molecular structure is greatly reduced. The simulation of a C_{60}^+ ion beam impact onto an ice film compared to the Au_3^+ ion impact, figure 2.2, shows that a much wider crater is formed with the C_{60}^+ ion beam. The C_{60}^+ ion beam impact ejects more material from the top layer of the target, with a reduced damage cascade (i.e. mixing) in the underlying Ag substrate. As the impact per ion is dramatically reduced, the use of the cluster ion beams allow larger fragments to be 'lifted' from the surface intact, thereby increasing their ion yields.

More recently, the ability to routinely perform depth profiling on a range of organic carbon based materials has got a step closer, with the advent of Ar_n^+

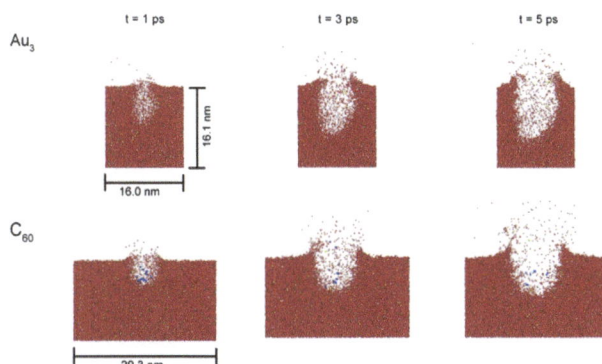

Figure 2.2. Simulation of 15 keV Au_3^+ and C_{60}^+ primary ion bombardment at a 40 degree angle of incidence on a 2.5 nm thick ice film (red) on Ag (blue) substrates. Each row of images shows the atom positions after 1, 3 and 5 ps. Reprinted from [13] with permission, copyright 2006, Elsevier.

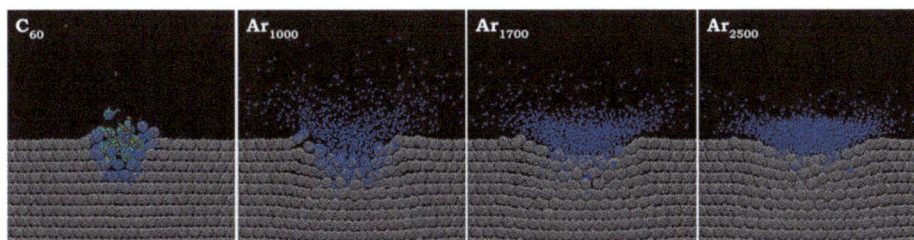

Figure 2.3. Simulation of C_{60}^+ and Ar_{18}, Ar_{1000}, Ar_{2500} ion beam impact on a fullerene substrate after 3 ps. Reprinted (adapted) from [19] with permission, copyright 2013, American Chemical Society.

$(500 \leqslant n \leqslant 2500)$ gas cluster ion beams (GCIBs). The very large clusters of argon used in the ion beams are formed by the supersonic expansion of high-pressure gas argon through a nozzle [14]. The evolution of the Ar_n^+ cluster ion beam source has enabled the depth profiling of polymeric and biological systems [15–18]. Similarly to the C_{60}^+ cluster ion beams, the very large cluster ion sources have extremely low impact energies per Ar^+ atom. Figure 2.3 shows the simulation of a C_{60}^+ ion beam impact compared to the impact of Ar_{18}, Ar_{1000}, Ar_{2500} clusters, into a fullerene-based material [19]. The simulation in figure 2.3 shows that the size and shape of the damaged area caused by the C_{60}^+ and smaller Ar_n^+ cluster ions are similar, with the energy due to the ion impacts deposited in the very near surface. As the argon cluster size is increased, the shape of the damaged region changes; the lateral size of the damage increases whilst its depth into the target decreases. This significant change in the way in which the projectile interacts with the target material leads to both an increase in the ejection of intact molecules from the sputtered region and reduced fragmentation of the sputtered molecules. The use of these large cluster ion beams does not however lead to an increase in secondary ion yields [20], but as much less fragmentation occurs and more of the parent molecules are captured intact, a cleaner mass spectrum is produced [21]. This is very important for the differentiation of materials where the chemical bonding

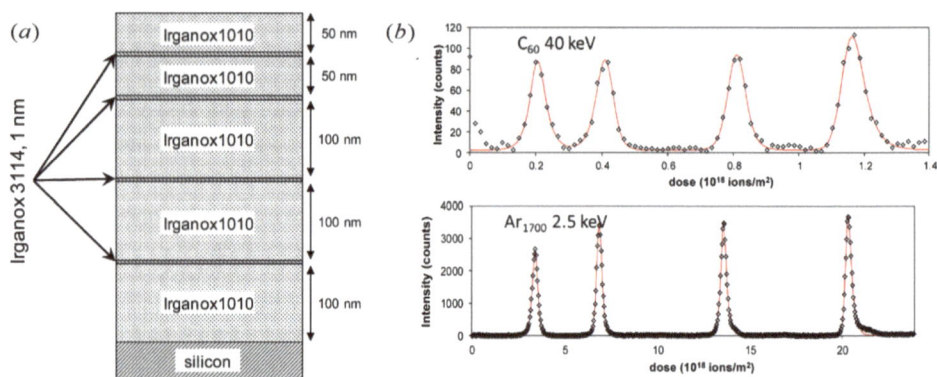

Figure 2.4. (*a*) Schematic of the structure of the multi-layered sample used in the interlaboratory study. (*b*) The depth profiles obtained using a C_{60}^+ and Ar_n^+ cluster ion source, plotting the [M3114–R]$^-$ secondary-ion intensities against sputtering ion dose. Reprinted (adapted) from [18] with permission, copyright 2012, American Chemical Society.

is the only difference in the sample (e.g. C–H bonded materials) as opposed to detecting elemental changes of composition.

Recently, an inter-laboratory study highlighted the performance of the argon cluster ion beam during depth profiling [18]. An organic multi-layer reference material made of a 400 nm thick Irganox 1010 matrix with 1 nm marker layers of Irganox 3114, placed at depths of 50, 100, 200 and 300 nm, was analysed. A schematic of the structure is shown in figure 2.4(*a*). Depth profiles were obtained using both C_{60}^+ and Ar_n^+ cluster ion sources, figure 2.4(*b*). The results showed that an improved, and constant, depth resolution was obtained with the Ar^+ cluster ion source compared to the C_{60}^+ ion source.

The depth profiling of organic materials with the Ar_n^+ cluster ion beam appears to have now superseded most depth profiling applications of the C_{60}^+ ion beam. Previous studies have shown that the sputter yield, Y, for an Ar_n^+ cluster ion beam is linearly dependent on E, the energy of the ion beam [22]. In contrast, Y was found to decrease for increasing values of n, the number of argon atoms in the cluster, for a 20 keV Ar_n^+ ion beam [23].

Using data available in the literature, Seah has been able to define a universal equation for argon gas cluster sputtering yields for eight materials; Au, SiO_2, Si, Irganox 1010, HTM-1 (a model OLED material), polystyrene, polycarbonate and polymethylmethacrylate (PMMA) [24]. The yield, Y, of atoms sputtered per primary ion can be expressed by the universal equation:

$$\frac{Y}{n} = \frac{\left(\dfrac{E}{An}\right)^q}{\left[1 + \left(\dfrac{E}{An}\right)^{q-1}\right]} \tag{2.1}$$

where the parameters A and q are obtained by fitting.

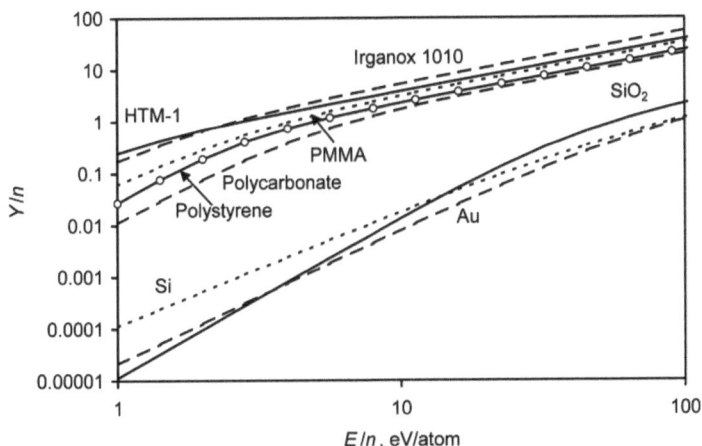

Figure 2.5. The compiled plot of Y/n for the eight materials analysed. The data were obtained at an incidence of 45° except for PMMA and Si which was obtained at 0° incidence. Reprinted from [23] with permission, Crown Copyright (2013).

The combined data for all the materials studied was plotted and is shown in figure 2.5. The plot shows two important points: firstly that the sputter rate for the organic materials is 2–4 times higher than for the elements or compounds such as SiO_2. The second important point is that over the range of energies per argon atom (E/n) investigated, the plot shows that the sputter rate of both the organic and inorganic sample can change by a factor of 100.

In the dual-beam ToF-SIMS systems the focus of both the C_{60}^+ and Ar_n^+ cluster ion beams is typically quite poor compared to the well-focussed LMIG sources. Consequently, secondary ion mapping of biological samples is still mostly carried out using the well-focussed LMIGs whilst depth profiling and the removal of material is carried out using the cluster ion beams.

The depth profiling of multilayer polymers and other organics with an Ar_n^+ cluster ion beam has been shown to be very promising; however, due to their novelty, the depth profiling of biological materials is currently not as common. As these cluster ion sources are fitted onto more and more instruments and become more widely available this area of research will no doubt grow very rapidly.

2.3 Mass analysis

Once the secondary ions have been generated they need to be separated according to their mass to charge ratio (m/z) in order to record the mass spectrum. There are several commonly used mass analysers; quadrupole, magnetic sector, time-of-flight, ion trap and orbitrap. Each of the analysers uses different physical principles to separate the ions, and as a consequence all have different upper mass limits, mass resolutions and mass accuracy [25]. Of these analysers, the most common and most simple is the ToF system, whereby the secondary ions are separated by their flight time in a flight tube without the use of an electric or magnetic field.

As the primary ion beam is rastered over the sample surface it is pulsed in extremely short pulses, which then creates a pulse of secondary ions from the sample

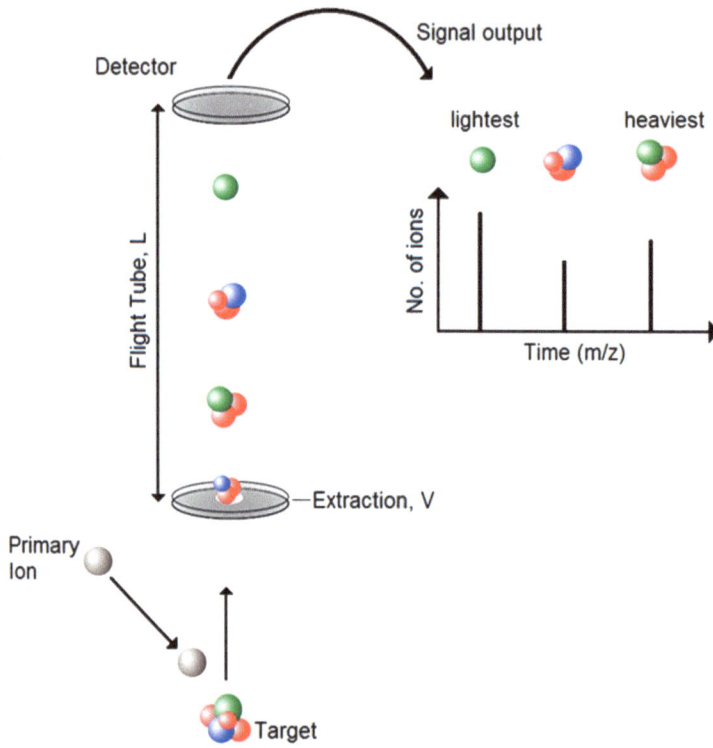

Figure 2.6. Schematic showing the time-of-flight of the sputtered secondary ions through the flight tube. The lightest secondary ion arrives first at the detector followed by the heavier secondary ions.

surface. These are accelerated through an extraction plate held at a fixed potential, V, typically between 2 and 8 KeV into the ToF analyser. All the secondary ions now have the same kinetic energy, and as $KE = \frac{1}{2}mv^2$ the velocity of the ions is now inversely dependent on the square root of their mass. After acceleration the ions drift through a field free tube (the flight tube), see figure 2.6, of known length, L, the mass separation (m/z) is obtained according to equation (2.2) below.

$$\frac{m}{z} = \frac{2Vt^2}{L^2} \qquad (2.2)$$

where V is the accelerating potential, L is the length of the flight tube, and t is the time taken for the ions to fly through the flight tube and strike the detector i.e. the time-of-flight of the secondary ions.

The lightest ions will travel the fastest and arrive first at the detector. The heavier ions will move more slowly and arrive at later intervals. The arrival times of the ions at the detector will be dependent on the ion mass, thus from equation (2.2), the arrival times can be transformed into the respective m/z ratio generating a mass spectrum from each pulse of the ion beam.

At higher masses resolution becomes more difficult as flight times are longer and not all of the ions with the same m/z reach their ideal ToF velocities. To remedy this

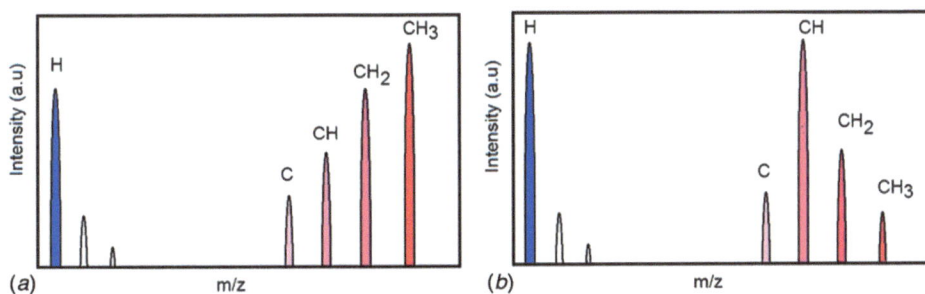

Figure 2.7. An example of the common peak pattern observed at the lower mass end of a mass spectra with (*a*) positive secondary ions and (*b*) negative secondary ions. These are used for initial mass calibration of the *x*-axis.

problem a series of high voltage ring electrodes are placed at the end of the analyser: a reflectron.

The reflectron improves the resolution at the higher masses by narrowing the range of flight times for a single *m/z* value. Faster ions will travel further into the reflectron compared to slower moving ions. Thus both fast and slow ions of the same *m/z* value will reach the detector at the same time, so narrowing the bandwidth of the output signals. For a commonly used reflectron-ToF mass analyser the upper mass limit that can be recorded has an *m/z* of 10 000. Typically mass resolution may be as good as 20 000 with an accuracy of 10 ppm, but these limits of detection are dependent on many factors related to individual instrumentation.

Mass calibration
Along with the simplicity of design, the time-of-flight mass analyser also has the advantage that it is self-calibrating. For the different polarity of secondary ions captured, positive or negative, two very distinct mass spectra are obtained at the lower mass range. Internal mass calibration can consequently be based on the light mass fragment ions that are always present in the spectra. The patterns for the positive and negative secondary ions are shown in figures 2.7(*a*) and (*b*) respectively. Most obviously, the first peak that is observed is hydrogen at (nominally) 1 amu. There then follows a 'packet' of peaks which are carbon and 3 hydrocarbon molecular peaks: CH, CH_2 and CH_3. The intensities of each of these peaks differ depending on the polarity of the secondary ions being detected. In the positive ion mode, CH_3^+ is the most intense with the preceding peaks dropping in intensity. In the negative ion mode, the CH^- is the most intense, with the CH_3^- ion now being the least intense, and may in some instances not be visible. Being able to recognise these two patterns at the lower mass end of the positive and negative ion mass spectra allows for the calibration of the *x*-axis from time to *m/z*.

For calibration further along the mass spectra, peaks can be selected for calibration. As the precision of time-to-digital converters is very high (these devices have a very good linearity over the full range), the relationship between $(m/z)^{1/2}$ and the ToF remains linear over the whole mass range. Along the spectra, distinctive masses can be relied upon for calibration, and again these are specific to the polarity

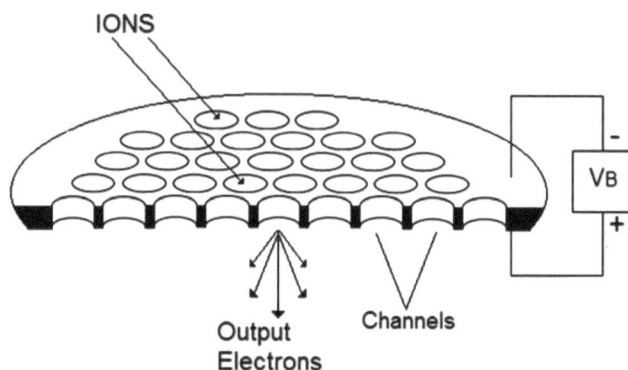

Figure 2.8. Schematic of a micro-channel plate used in the detection of secondary ions.

of secondary ions being collected. For positive mass spectra the formula $C_{2x-1}H_x{}^+$ can be used, with peak intensities typically higher for odd H ion content. For negative secondary ion mass spectra, the C_x^- and/or C_xH^- series may be used, along with the very electronegative elemental ions such as F^- and Cl^-.

2.4 Ion detection

After passing through the mass analyser the secondary ions are detected and amplified by a microchannel plate (MCP). MCPs combine several unique properties such as high gain, high spatial resolution and high temporal resolution making them useful for many applications. A MCP is a two dimensional array of very small glass channels (~8 μm) that are fused together and then sliced into a plate, figure 2.8. A single ion will enter a channel, strike the channel wall, and emit an electron. As there are many channels in the plate the process is continuously repeated creating a cascade of several thousands of electrons to emerge from the back of the plate. More than one MCP may be operated in series, leading to an amplification of one single event into an output of ~10^8 electrons.

Detectors that are based on MCPs can have a variety of designs, optimised for their particular use, but they will generally consist of three parts:
   (1) A converter—a mechanism to convert the initial particles into electrons.
   (2) An assembly of MCPs—a mechanism to amplify the initial single electrons into an electron pulse.
   (3) A readout device—a mechanism to detect the electron avalanche.

The final design of the detector will depend on the type of particles being detected, the time and position resolution, imaging area, linearity and sensitivity, signal-to-noise ratio, and signal throughput (counts/second).

2.5 Ultra high vacuum

Along with the development of ion beam sources and faster electronics for better counting statistics, many of the advances and improvements in sensitivity in mass spectrometry have occurred due to the improvement in the vacuum environment. Although this is not strictly a part of the analytical technique the importance of the vacuum cannot and should not be underestimated, as it is a key ingredient needed for any high sensitivity surface science experiments.

The vacuum in a modern ToF-SIMS analysis chamber will be as low as 10^{-10} mbar. This is obtained through an arrangement of roughing pumps, turbo molecular pumps, ion pumps and sublimation gettering pumps. Such low vacuum pressures are essential for two reasons: firstly, to maintain a surface that will stay clean for the duration of the surface science experiment, and secondly, to maximise the inelastic mean free path that the primary ion beam generated secondary ions can traverse from the analysis chamber into the analyser, without encountering losses due to gas phase scattering phenomena.

From the kinetic theory of gases we can gain an estimate of the number of gas molecules (the flux F) arriving at a surface and the time, therefore, for a monolayer of this gas to cover the surface. The incident flux of gas molecules is given by the equations:

$$F = \frac{P}{\sqrt{(2\pi M k_b T)}} \quad (2.3)$$

where P is the pressure (Nm^{-2}), M is the mass (kg), T the temperature (K) and k_b the Boltzmann constant. The time taken for a surface to be covered by a monolayer of adsorbate (e.g. dry air) is estimated by assuming a unit sticking probability of 1 and a coverage of 10^{19} per m^2, giving:

$$\text{Time (s)} = \frac{10^{19}}{F}. \quad (2.4)$$

Table 2.1 shows how improving the vacuum dramatically alters the time taken for adsorbate to form a monolayer on a surface. At the very high vacuums used in SIMS instrumentation, the time taken is almost 10 h, compared to only 3.5 s at 10^{-6} mbar, the vacuum pressure typically found in SEM instrumentation.

Table 2.1. The amount of time taken for a monolayer (ML) of adsorbate (dry air) to form on a surface at varying vacuum pressures.

Degree of vacuum	Pressure (Mbar)	Flux (m^{-2} s^{-1})	Time (ML) (s)
Atmospheric	1000	2.8×10^{27}	3.5×10^{-9}
Low	1	2.8×10^{24}	3.5×10^{-6}
Medium	10^{-3}	2.8×10^{21}	3.5×10^{-3}
High	10^{-6}	2.8×10^{18}	3.5
Very High	10^{-9}	2.8×10^{18}	3.5×10^{3}
Ultra High	10^{-10}	2.8×10^{14}	3.5×10^{4}

References

[1] Winters H F and Coburn J W 1976 *Appl. Phys. Lett.* **28** 176
[2] Bolbach G, Viari A, Galera R, Brunot A and Blais J C 1992 *Int. J. Mass Spectrom. Ion Process.* **112** 9
[3] Shard A G, Spencer S J, Smith S A, Havelund R and Gilmore I S *Int. J. Mass Spectrom.* **377** 599–609
[4] Touboul D, Halgand F, Brunelle A, Kersting R, Tallarek E, Hagenhoof B and Laprevote O 2004 *Anal. Chem.* **76** 1550–59
[5] Kollmer F 2004 *Appl. Surf. Sci.* **231–232** 153
[6] Brunelle A, Touboul D and Laprevote O 2005 *J. Mass Spectrom.* **40** 985–99
[7] Jones E A, Fletcher S J, Thompson C E, Jackson D A, Lockyer N P and Vickerman J C 2006 *Appl. Surf. Sci.* **252** 6844–54
[8] Jones E A, Lockyer N P and Vickerman J C 2007 *Int. J. Mass Spectrom.* **260** 146–57
[9] Postawa Z, Czerwinski B, Szewczyk M, Smiley E J, Winograd N and Garrison B J 2003 *Anal. Chem.* **75** 4402
[10] Postawa Z, Czerwinski B, Szewczyk M, Smiley E J, Winograd N and Garrison B J 2004 *J. Phys. Chem.* B **108** 7831
[11] Delcorte A, Wehbe N, Bertrand P and Garrison B 2008 *Appl. Surf. Sci.* **255** 1229
[12] Declorte A, Garrision B J and Hamraoui K 2009 *Anal. Chem.* **81** 6676–86
[13] Russo Jr M F *et al* 2006 *Appl. Surf. Sci.* **252** 6423–5
[14] Matsuo J, Okubo C, Seki T, Aoki T, Toyoda N and Yamada I 2004 *Nucl. Instrum. Methods Phys. Res.* B **219–220** 463–7
[15] Bich C, Havelund R, Moellers R, Touboul D, Kollmer F, Niehuis E, Gilmore I S and Brunelle A 2013 *Anal. Chem.* **85** 7745–52
[16] Mouhib T, Poleuins C, Wehbe N, Michels J J, Galagan Y, Houssiau L, Bertrand P and Delcorte A 2013 *Analyst* **138** 6801–10
[17] Aoyagi S, Fletcher J S, Sheraz (Rabbani) S, Kawashima T, Berrueta-Razo I, Henderson A, Lockyer N P and Vickerman J C 2013 *Anal. Bioanal. Chem.* **405** 6621–8
[18] Shard A G *et al* 2012 *Anal. Chem.* **84** 7865–73
[19] Czerwinski B and Delcorte A 2013 *J. Phys. Chem.* C **117** 3595–604
[20] Postawa Z, Paruch R, Rzeznik L and Garrison B J 2013 *Surf. Int. Anal.* **45** 35–38
[21] Kayser S, Rading D, Moellers R, Kollmer F and Niehuis 2013 *Surf. Int. Anal.* **45** 131–3
[22] Matsuo J, Toyoda N, Akizuki M and Yamada I 1997 *Nucl. Instrum. Methods Phys. Res.* B **121** 459–63
[23] Seki T, Murase T and Matsuo J 2006 *Nucl. Instrum. Methods Phys. Res.* B **242** 179–81
[24] Seah M 2013 *J. Phys. Chem.* C **117** 12622–32
[25] De Hoffmann E, Charette J and Stroobant V 2007 *Mass Spectrometry: Principles and Applications* (Chichester: Wiley)

IOP Concise Physics

An Introduction to Time-of-Flight Secondary Ion
Mass Spectrometry (ToF-SIMS) and its Application
to Materials Science

Sarah Fearn

Chapter 3

Modes of analysis

Dual-beam ToF-SIMS instruments can be operated in different modes to give different analysis outputs, depending on how the primary and sputter ion beams are used in conjunction with one another. The different types of analyses that can be obtained are: high-resolution mass spectra, ion images (chemical maps) and depth profiles (and 3D analysis).

3.1 High-resolution mass spectra

High-resolution mass spectra are obtained from sample surfaces by sole use of the primary ion beam. Very short primary ion beam pulses (~nS) are used to irradiate a pre-defined spot, figure 3.1(a). The generated secondary ions are then extracted and accelerated into the time-of-flight analyser, figure 3.1(b). The secondary ions are then separated according to their mass-to-charge ratio (m/z). The mass spectrum is recorded along with the ion beam spot co-ordinates. The primary ion beam is then moved to an adjacent pixel and the spot is irradiated by the short ion beam pulse. The process is repeated until the desired area (x, y) has been scanned and analysed, producing a mass spectra, figures 3.1(c) and (d). Due to the very short primary ion beam pulses the mass resolution of the mass spectra is very high and typically a mass resolution, ($M/\Delta M$), of 10 000 is routinely obtained.

During the analyses the ion beam dose must be strictly controlled to ensure that the surface is not sputtered away. This is known as 'static SIMS'. Static SIMS uses a primary ion beam dose which minimises the interaction of the primary ion beam with the top layer of atoms or molecules, such that less than 1% of the surface is removed [1]. This limits, therefore, the ion beam dose that can be used during an analysis to approximately 10^{15} ions cm^{-2} for inorganic materials. However, for polymer and biological materials the static limit will be much lower and dependent on the structure

doi:10.1088/978-1-6817-4088-1ch3

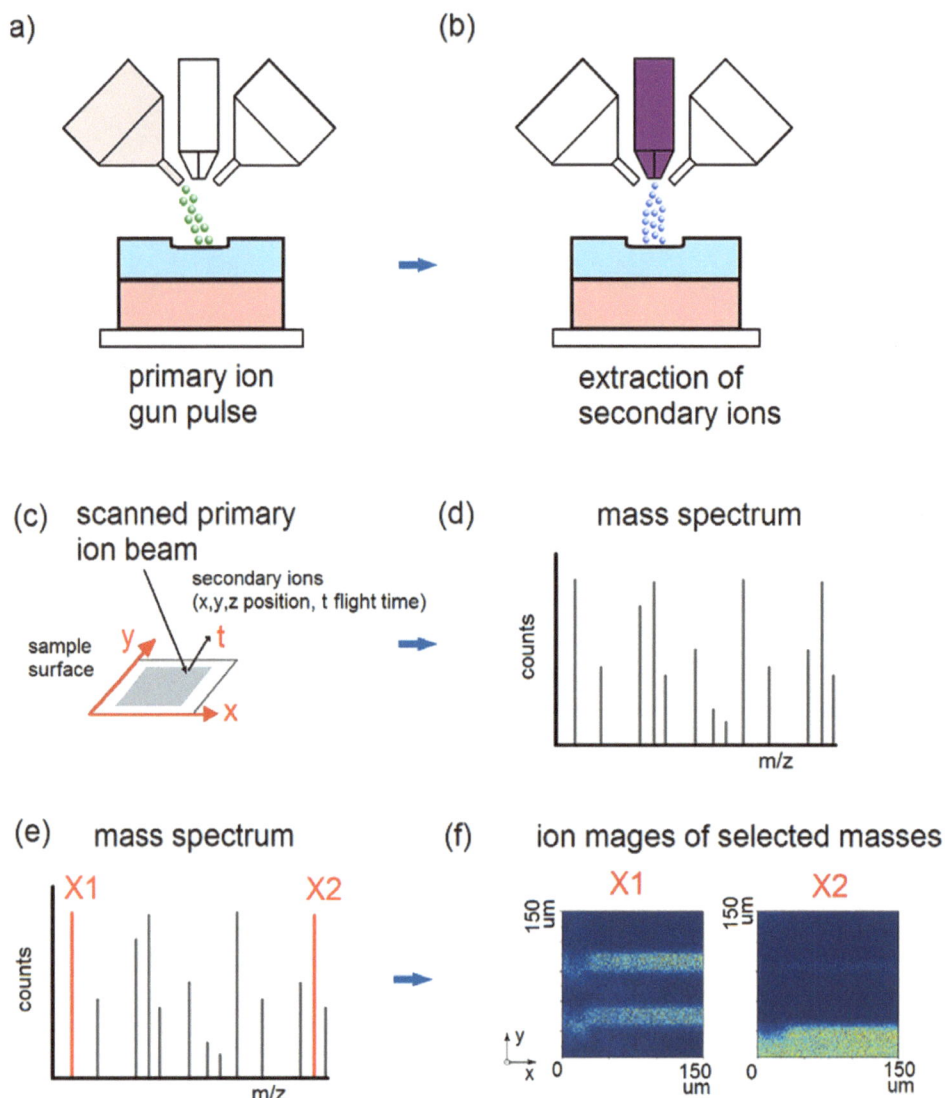

Figure 3.1. The steps needed to obtain mass spectra and ion images. (*a*) A short pulse from the primary ion beam irradiates a spot and sputters secondary ions. (*b*) These are collected in the analyser, and separated according to (*m/z*). (*c*) The process is repeated on the neighbouring pixel, until a specified area has been analysed. (*d*) Mass spectra are produced as the detector collects the sputtered secondary ions. (*e*) By selecting specific peaks from the mass spectrum, (*f*) the spatial distribution of the selected peaks can be generated to give ion images as the spatial co-ordinates for the secondary ion signals are known.

of the surface being analysed and will vary from 10^{10} to 10^{13} ions cm^{-2}. The application of such low ion beam doses and the reduced ion beam interaction with the surface effectively renders the analysis non-destructive. The use of such low ion beam doses does have consequences for the amount of secondary ions that are subsequently created for detection. As the sample is left undamaged it means that it can still be used for other characterisation techniques.

During the mass spectra analysis, the spatial co-ordinates of each irradiated pixel are recorded. Specific ions may be selected from the mass spectra coupled with their distribution (i.e. their location) over the defined analysed area (x, y). These are known as ion images or chemical ion maps, figures 3.1(e) and (f). The size of the analytical area can be varied widely and may be set as small as 5 μm^2 and up to areas of several millimetres. Depending on the pulse time of the primary ion beam, the lateral resolution can reach up to ~250 nm.

3.2 Depth profiling

For many systems, it is desirable to gain an understanding of how composition varies with depth, for example, the diffusion of elements from implants into bone and vice versa, or how well defined an interface may be. Through the use of an additional ion source known as a sputter ion beam/gun it is possible to depth profile through the sample and thereby measure mass spectra as a function of depth. In this instance the ion beam dose exceeds the static limit, and layers of material are 'peeled away' by the sputter gun to expose further layers of atoms or molecules for analysis. As the ion beam dose increases, more and more material is removed forming a sputter crater. This type of analysis is known as 'dynamic SIMS' and is much more destructive compared to static SIMS and ion beam doses greatly exceed the static limit. This type of analysis is typically carried out to characterise interfaces, compositional changes through a material due to a processing step, and features beyond the surface region [2].

The three key steps of depth profiling in a dual-beam ToF-SIMS are outlined in figure 3.2. The first step in depth profiling is to raster the sputter beam over the pre-defined sputter area (typically a square with side length between 250 and 500 μm), figure 3.2(a). The sputter ion beam is generally of a much higher current than the analytical ion beam, typically on the order of tens to hundreds of nAs. The higher ion beam dose results in the removal of material in a controlled fashion, and the current must be stable to maintain a constant sputter rate during the analysis. It is also worthwhile to highlight that in addition to removing a large amount of material the sputter beam may cause the surface chemistry of the sample to be altered affecting the ion yields, hence careful selection of the sputter beam is essential to optimise the accuracy of measurement. After sputtering the primary ion beam is applied to a smaller area, or sub-crater, within the larger sputter crater, figure 3.2(b), causing the emission of secondary ions as described in the previous section. A sub-crater is utilised to minimise any interference from material on the edge of the sputter crater, referred to as the 'edge effect'. After the secondary ions have been collected in the analyser, the sputtering and primary ion beam analysis is repeated sequentially until the desired depth of sample has been measured. Again the detected secondary ions produce a mass spectrum. Peaks of interest are selected from the mass spectra, figure 3.2(d), and a depth profile of those species is generated,

(a)

(b)

(c)

sputter ion gun

primary ion
gun pulse

extraction of
secondary ions

(d)   mass spectrum

counts

m/z

peaks are selected from the mass
spectrum to form a depth profile

(e)

intensity

depth / nm

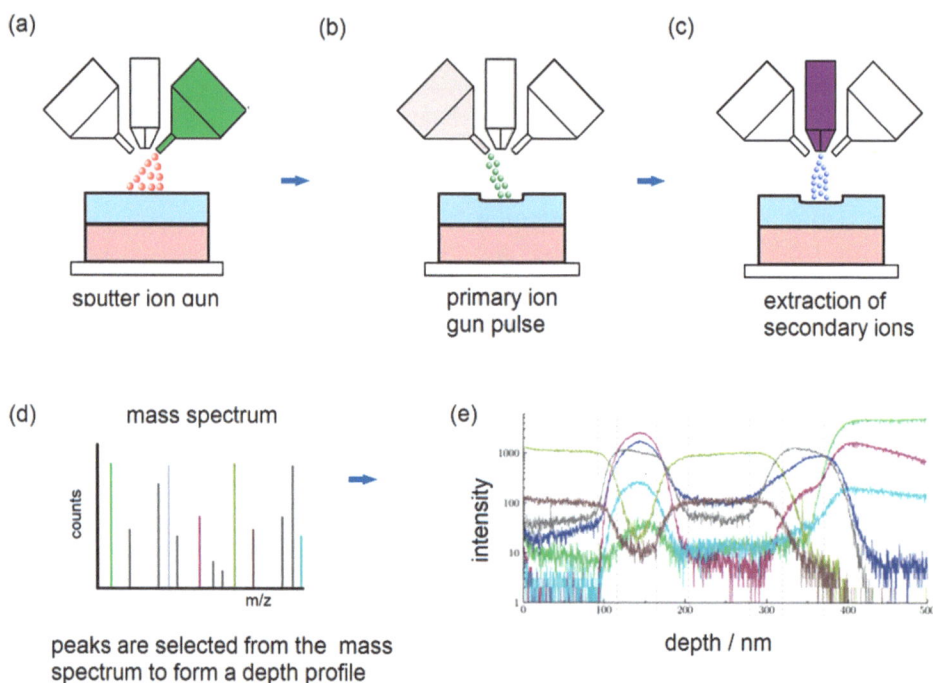

Figure 3.2. Schematic representation of the three main steps used in depth profile ToF-SIMS. (*a*) A high-energy sputter beam is rastered over the sputter area. (*b*) The analytical beam is applied to the analysis area which is centred within the sputter area. (*c*) The secondary ions generated by the analysis beam are collected and accelerated into the analyser, and a mass spectrum produced. (*d*) By selecting certain peaks a depth profile of the species can be obtained, (*e*).

figure 3.2(*e*). Finally, to avoid any charge build up that may occur on an insulating sample a flood of low energy electrons can be irradiated onto the surface.

As previously mentioned, the location of each generated mass spectra, and thus the selected peaks, are known and 2D ion maps can be generated. This also applies to chemical information obtained during depth profiling, but now also includes the z-axes. As the detected secondary ions are now registered on the three axes 3D chemical plots can be constructed. Figure 3.3(*a*) shows how during the analyses all the detected secondary ions as a function of the cube (x, y, z) co-ordinates are stored. After the analysis it is possible to reconstruct the secondary ion images from different depths: the x–y planes, as well as through the cube: the x–z and y–z planes. Figure 3.3(*b*) shows the total secondary ion images for H and S on the x–y plane. The green lines on the image indicate where slices were taken along the x–z and y–z planes, figures 3.3(*c*) and (*d*) respectively. The slices through the cube show the different distribution of hydrogen and sulphur in the layered structure. In particular layer heterogeneities can be seen. In the total ion image for the x–y plane a feature is observed, highlighted by the red circle. By extracting the y–z plane ion image it can be seen that the feature (a pinhole) runs all the way through the analysed structure.

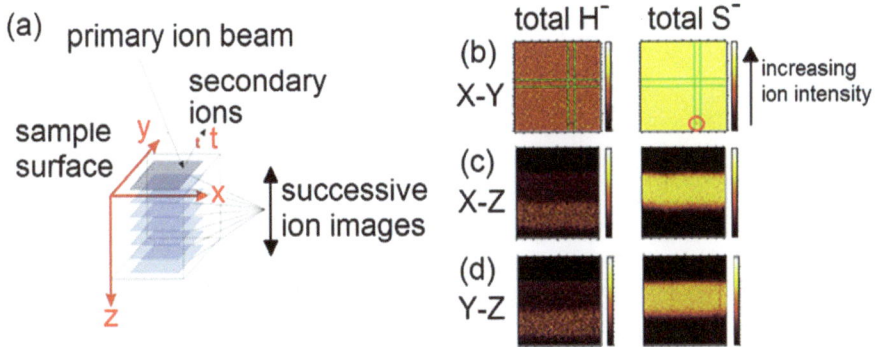

Figure 3.3. (*a*) How the secondary ion data is registered to its spatial co-ordinates, in all three dimensions. (*b*) The total ion counts on the *x*–*y* plane for H and S respectively. (*c*) The H and S ion map for a slice in the *x*–*z* plane and (*d*) the H and S ion maps for a slice in the *y*–*z* plane.

References

[1] Briggs D and Hearn M J 1986 *Vacuum* **36** 1005
[2] Wilson R G, Stevie F A and Magee C W 1989 *Secondary Ion Mass Spectrometry: a Practical Handbook for Depth Profiling and Bulk Impurity Analysis* (New York: Wiley)

An Introduction to Time-of-Flight Secondary Ion
Mass Spectrometry (ToF-SIMS) and its Application
to Materials Science

Sarah Fearn

Chapter 4

Ion beam–target interactions

The bombardment of a target with an ion beam causes both the removal and
ionisation of the target atoms along with a number of deleterious beam–target
interactions shown schematically in figure 4.1 [1]. The deleterious ion beam–target
interactions can result in distorting the analyses attained during a SIMS experiment.
These interactions relate primarily to SIMS experiments whereby the ion beam dose
exceeds the static limit and dynamic SIMS is being carried out (i.e. depth profiling). It
is important to be aware of these deleterious ion beam processes and how they present
themselves in the resulting data so that incorrect interpretation of data is not made.

The two major interactions between the ion beam and the target atoms are
recoil implantation and cascade mixing. The ion beam hits the target surface
and penetrates into the near surface. The bombarding ions possess energy and
momentum, which they share with the target atoms. Recoil implantation (number 1,
figure 4.1) occurs through direct impact collisions between the ions and the target
atoms. Only enough energy is given to the target atoms to knock them anisotropically
into the solid, where they come to rest [2]. Cascade mixing (number 2, figure 4.1)
occurs as the interactions between the primary ions and the resident atoms cause
further collisions creating an isotropic collision cascade. The cascade continues to
develop until the transferable energies become less than the displacement energy
of the target atoms [3]. In both cases any excess energy is released as phonons. The
primary beam species comes to rest at various depths within the host lattice. The
penetration depth of these ions is dependent on the beam energy and angle of
incidence of the primary beam with respect to the target surface.

An atom or group of atoms may receive enough energy in a suitable direction
enabling them to overcome the surface binding forces and be sputtered from the
target (number 4, figure 4.1). The approximate depth of origin of these emitted

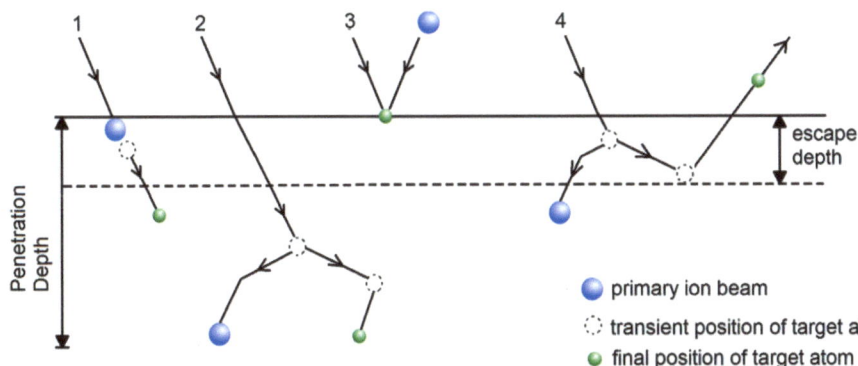

Figure 4.1. Schematic of the beam–target interactions occurring during SIMS analysis: 1 recoil implantation, 2 cascade mixing, 3 primary ion reflection, 4 cascade mixing with sputtering.

atoms and ions is ~2–3 atom layers for inorganic crystalline materials [4], but as mentioned earlier the location of the emitted particles can be up to 10 nm away from the initial impact of the primary ion [5]. The potentially small escape depth of the secondary ion signal is further degraded due to the deleterious ion beam–target interactions. Some of these effects will be discussed briefly in the following section, but more detailed information can be found in the literature [2, 6].

4.1 Ion beam induced atomic mixing

Cascade mixing and recoil implantation are the most well-known factors limiting depth resolution, giving rise to broadening of buried features and interfaces. Out of the two processes it has been suggested that cascade mixing is the dominant material transport process accompanying sputtering [7–9].

The collision cascade initiated by the primary ion impact is capable of efficiently mixing surface and sub-surface layers over a distance on the order of the primary ion range, R_p [10]. The efficiency of the cascade mixing is such that the mixed zone can propagate through the material faster than the sputtering erodes the surface. At this point in a SIMS analysis, the ion beam is essentially now analysing a mixed layer which is composed of the original host material and the implanted ion beam species [8]. The mixed layer is continuously pushed into the substrate ahead of the sputtering front resulting in final depth profiles that decay exponentially resulting in asymmetric depth profiles [11]. The influence of cascade mixing on SIMS depth profiles can, however, be reduced by lowering the energy of the primary beam, by the use of heavy ions or by the use of glancing angles (or some combination of these factors).

4.2 Beam induced surface roughening and uneven etching

Another major influence on the SIMS profiles is the effect of surface roughening and uneven sputtering. The evolution of topography and uneven etching are observed in the ion beam sputtered region due to non-uniform sputtering. The effects of roughening are particularly evident on multilayer structures. The depth resolution typically degrades with depth, and the angle of incidence of the ion beam can also

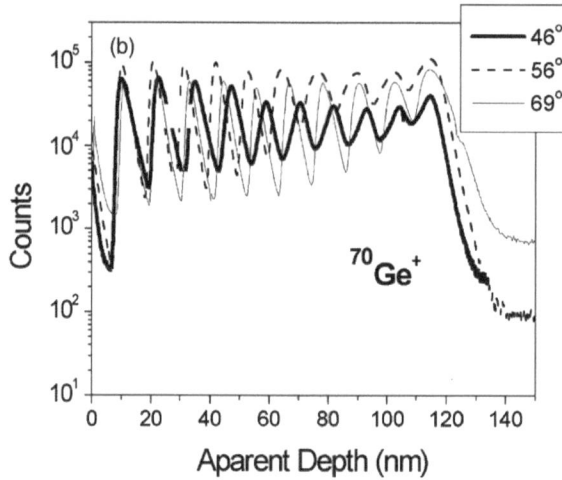

Figure 4.2. A SIMS depth profile through a SiGe δ-doped sample. The onset of roughening and loss of depth resolution occurs at different depths for the different angles of ion beam incidence on the sample. Reprinted from [12] with permission, copyright 2003, Elsevier.

have an influence, as highlighted in figure 4.2 which shows a depth profile through a SiGe δ-doped sample. As the sample surface roughens, the SIMS depth profile through a multi-layered structure will display increasingly broad peaks [12]. The peak to valley dynamic range also degrades.

Induced topography, such as cone formation, has been known about for some time [13, 14]. The phenomena has been extensively investigated and found to be ion beam and sample dependent, that is to say, the topographical roughness of a large grained metal will be different to that of an amorphous layer [15].

Topography can also develop due to preferential sputtering. In this case one element is sputtered away more rapidly than another. The extent of preferential sputtering is influenced by several factors including crystallisation and chemical properties. Preferential sputtering will occur along edge features and grain boundaries, leading to the formation of cones in the sputtered region [14].

In the case of uneven etching, adjacent layers may be exposed to the ion beam at the same time, producing an unresolvable depth profile. Using various SIMS instruments, the influence of uneven etching was empirically and theoretically studied [16]. It was found that as little as 0.1% unevenness of the crater base could influence a depth profile. A periodic doping multi-layer structure was modelled with 1 and 10% unevenness on the crater base. The results showed that the inter-peak dynamic range decreased with depth, and was bounded by amplitude envelopes. In the case of the crater with 1% unevenness, the envelopes met after 100 peaks whereas for the crater with 10% unevenness they met at the end of 10 peaks, with the 10th and 11th layers being unresolvable.

4.3 Beam induced segregation

The use of certain reactive ion beams for SIMS analyses can improve ionisation efficiencies of the target material and therefore improve the analysis. During SIMS depth profiling with reactive ion beams such as oxygen, however, beam induced segregation can occur. The phenomenon is identified by SIMS profiles with grossly extended decay lengths.

Various suggestions have been put forward to explain why the segregation occurs, but explanations based on thermodynamics and field induced segregation have emerged as the most consistent mechanisms to explain the process.

The thermodynamic argument, or Gibbsian segregation [17], depends on the relative heat of formation of the compound formed during reactive sputter profiling. The process can be most easily explained by considering copper implanted silicon under oxygen bombardment at normal incidence. Initially upon sputtering, an altered surface layer of SiO_2 is formed as the beam reacts with the sample. The relative heat of formation of SiO_2 is -217 kcal mol^{-1}, whereas for CuO and CuO_2 it is -37 and -40 kcal mol^{-1} respectively [18]. Under oxygen bombardment, therefore, it is more energetically favourable for SiO_2 to be formed and the Cu will segregate away from the beam induced oxide. The Cu is not sputtered efficiently and the decay length for the impurity becomes grossly extended. Elements such as Ag, Pb and Au were also found to have similar segregation tendencies in silicon [19]. Investigating beam induced segregation effects caused by oxygen and nitrogen primary beam bombardment, Homma *et al* [20] found that in some cases the reverse process 'improved' decay lengths. In situations where it was more energetically favourable for the impurity to oxidise rather than the silicon, then the impurity would segregate toward the beam induced oxide and be more efficiently sputtered, leading to shorter decay lengths.

The field induced segregation model also involves the altered layer that is formed during reactive sputter profiling. Using silicon under oxygen bombardment again as an example, the SiO_2 altered layer that is formed on the sample surface is insulating. Continued ion bombardment causes the surface to become charged, and an electric field is induced across the insulating layer. The effect of the induced field on impurity segregation was observed in the differing decay lengths obtained for Cu implants in n- and p-type silicon [21, 22]. The Cu was found to segregate more in the n-type silicon than in than in the p-type material. This was explained by the fact that the charge at the surface of the altered layer differs in strength for the two types of silicon; due to the presence of a depletion layer in the p-type silicon, the surface charge was found to be higher [23]. Thus, as the field across the altered layer decreases, the segregation effects increase. This process was highlighted even more when analyses performed under electron bombardment, to eradicate surface charging, produced even longer decay lengths. By performing analyses with an O^- primary beam, the segregation effects were found to diminish [22]. Increased segregation effects under electron bombardment were also observed on As implanted Si [24].

The influence of the thermodynamic properties on the segregation effect is found to be important only for the location of the impurity relative to the Si/SiO_2 interface. The actual magnitude of the segregation is thought to be dictated by the strength of the induced electric field across the altered layer [22].

As the segregation effects are induced due to the formation of an oxide layer on the surface of the sample, one way of reducing the problems is to perform the analyses at angles so that the altered layer cannot form [20, 25]. By increasing the angle to past 30° incidence, Kilner *et al* found that the decay length of a Cu implant in Si could be reduced from 3000 nm to approximately 32 nm [25].

4.4 Other beam induced effects

Radiation enhanced diffusion and crystallinity of the matrix can also play a role in degrading the depth resolution. These factors are minor in comparison with the degrading processes mentioned previously, but can become dominant in certain situations.

Radiation enhanced diffusion is dependent on vacancies and other lattice defects generated by ion bombardment [26]. Profile broadening effects due to radiation enhanced diffusion were observed by Vandervorst *et al* in the redistribution of 5 keV As implants in Si subject to oxygen bombardment [27]. During high energy analysis it was found that collisional and recoil mixing did not account for the total broadening observed. It was concluded that the dominant broadening mechanism was radiation enhanced diffusion when there was no oxide present.

The effect of crystallinity on depth resolution was investigated by Hosler *et al* [28]. Ion channelling was found to impose the severest limitations on the depth resolution obtainable in the depth profiling of polycrystalline aluminium layers. The depth resolution could be improved, however, through an appropriate choice of ion species, ion energy and incidence angle.

4.5 Depth profiling with cluster ion beams

The application of cluster ion beams to the analysis of C-based materials has expanded the types of material that can be analysed in SIMS, however, they also have their own deleterious ion beam–target interactions. Upon impacting a C-based material with an energetic cluster ion beam, both fragmentation of the cluster ion beam and target material occurs. These processes are intensified during depth profiling experiments, and the profiles are typified by the rapid loss or drop off of the characteristic signal being monitored. During the bombardment with a C_{60}^+ cluster ion beam, the loss of secondary ion signal has been attributed to either cross-linking with the target material or carbonisation of the target surface from the C_{60} ion beam. Cross-linking occurs due to the target material forming newly bonded material within the molecules of the target as the ion beam dose increases, leading to a reduced sputtering efficiency. The second process, carbonisation, is the formation of an amorphous carbon layer on the target surface, which then blocks any further emission of secondary ions from the material below. The carbonisation process is a combination of both cross-linking and dehydrogenation of the organic sample [29].

The effect of cross-linking due to C_{60}^+ ion beam bombardment occurs readily in polystyrene. The loss of secondary ion signal during depth profiling is highlighted in figure 4.3(*a*) which shows a depth profile of a polystyrene (PS) layer on a silicon substrate, the PS layer is represented by the $C_7H_7^+$ and $C_{15}H_{13}^+$ secondary ion signals [30]. With increasing C_{60} ion beam dose the selected secondary ion signals rapidly drop off. The cause of the cross-linking has been related to the presence of free radicals which are formed in the target as bonds are broken during bombardment [31].

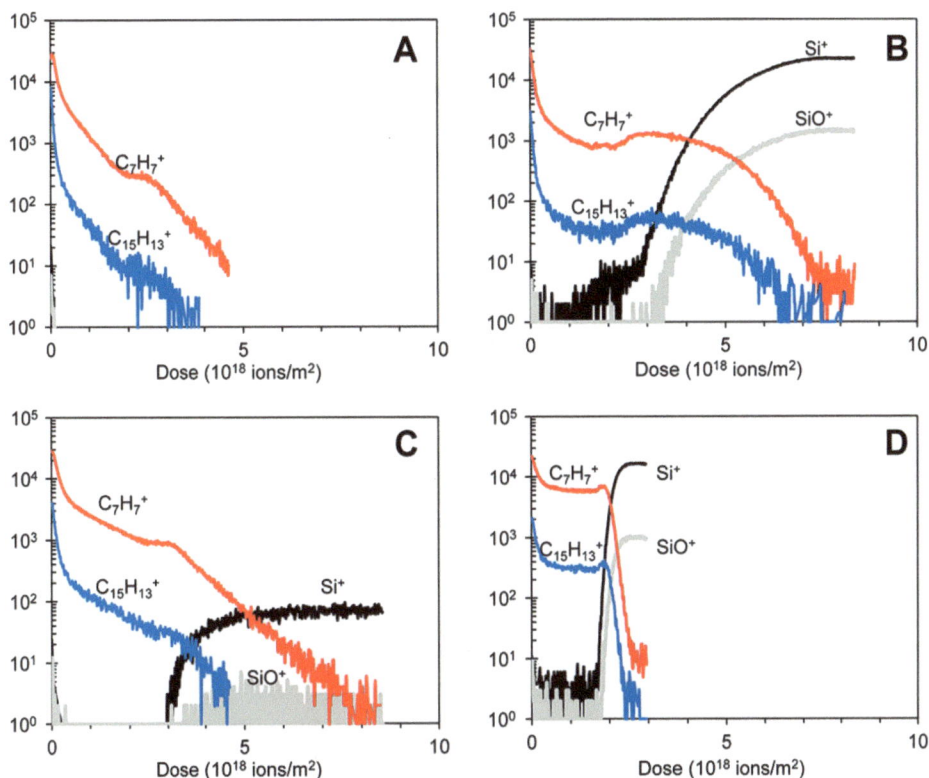

Figure 4.3. SIMS depth profiles from a PS layer on a silicon substrate acquired with a C_{60}^+ ion beam under four different conditions. (*a*) At 25 °C without NO dosing, (*b*) at 25 °C and 1×10^{-5} mbar NO, (*c*) at −100 °C without NO dosing, and (*d*) at −100 °C, 1×10^{-5} mbar NO. Secondary ion signals are Si^+ and SiO^+ for the substrate and $C_7H_7^+$ and $C_{15}H_{13}^+$ for the polystyrene. Reprinted (adapted) from [30] with permission, copyright 2013, American Chemical Society.

The C_{60} projectile also provides highly reactive carbon radicals which also interact with the target leading to the creation of additional cross-links [32].

The detrimental effects of the ion bombardment induced cross-linking can be somewhat off-set by flooding with nitrous oxide (NO), figure 4.3(*b*), and reducing the temperature of the sample, figure 4.3(*c*). By combining both a reduction in sample temperature and NO flooding the depth profile is dramatically improved, figure 4.3(*d*), however, NO flooding and reduced sample temperature does not work for other organic materials. In comparison to the Ar_n cluster ion guns, for very small Ar clusters ($n = 18$) similar effects related to cross-linking are observed. If the number of atoms in the Ar_n cluster is increased to values typically used with Ar cluster ion guns (500–5000), the crosslinking of molecules completely disappears.

As previously mentioned, although Ar_n^+ cluster ion beams are not currently universally available, initial studies, both experimentally and theoretically, indicate that they are much better than the C_{60}^+ cluster beams for depth profiling C-based materials. During the VAMAS study (previously discussed in section 2.2) the improvement in depth profiling using the Ar_n^+ GCIB was clearly displayed, however, as well as investigating the traditional effects such as roughening, etc, as discussed previously which typically relate to inorganic materials, there are other effects that need to be considered with relation to depth profiling C-based materials with Ar GCIBs.

The mechanism of ion formation from organic materials is poorly understood, with only theoretical models having been put forward in the past, without the ability to experimentally confirm the models being proposed. With the development of the Ar GCIB, which cause less ion beam impact damage coupled with well defined samples such as the Irganox 1010–3114 multi-layered structure, a clearer idea of where the secondary ion signals emanate from is possible. It is also, therefore, possible to study ion beam matrix effects that occur during the depth profiling of these well ordered materials [33].

Secondary ion enhancement and suppression have been clearly demonstrated in the depth profiling of thin markers layers of Irganox 3114 and Fmoc-PFLPA in Irganox 1010, and aluminium tris-(8-hydroxyquinolate) (Alq$_3$) in 4,4-bis[N-(1-napthyl-1-)-N-phenyl-amino]-biphenyl (NPB) with a 5 keV Ar$_{2000}^+$ ion beam. For a small fraction of Irganox 3114 in the Irganox 1010 no significant suppression or enhancement of the (M$_{1010}$-H)$^-$ ion was observed, figure 4.4(a), showing ideal behaviour. However, the presence of the Fmoc and Alq$_3$ layer causes a clear suppression and enhancement of the (M$_{1010}$-H)$^-$ ion, figure 4.4(b), and the C$_{44}$N$_2$H$_{22}^+$ ion, figure 4.4(c), respectively. The profiles of figures 4.4(b) and (c) indicate that for these mixtures, the secondary ion intensity is not proportional to the amount of material, and is no longer ideal, whereas no matrix effect is observed in figure 4.4(a).

Uniform binary mixtures of Irganox 1010 and Irganox 1098 have also been depth profiled with the argon cluster ion beam, and were used to identify further the matrix effects related to the suppression and enhancement of specific secondary ion signals. The degree of non-ideal ion intensity with composition was expressed as Ξ, the matrix effect magnitude: for secondary ion suppression Ξ is negative and for ion enhancement it is positive. By plotting the selected secondary ions from each of the polymers against the volume fraction of the mixture, it was possible to show how strong, and the sign, of the matrix effect magnitude for the selected secondary ions [33]. These well-defined materials may be the beginning of establishing a useful set of reference materials for future (and possibly quantifiable) organic depth profiling,

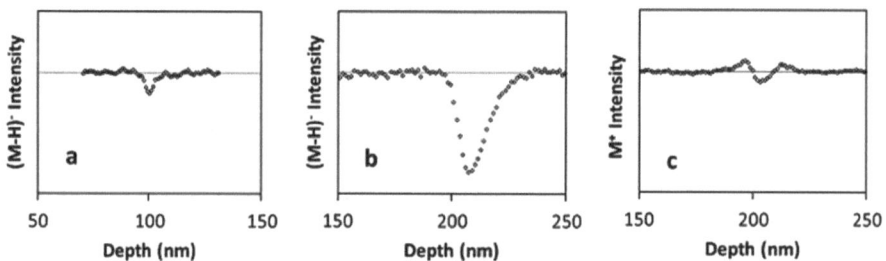

Figure 4.4. Secondary ion intensities for a matrix of (a) Irganox 1010 with a 1.0 nm layer of Irganox 3114, (b) and a 2.9 nm layer of Fmoc-PFLPA and (c) a 3.2 nm layer of Alq$_3$ in NPB. Plots (a) and (b) show the (M$_{1010}$-H)$^-$ intensity and (c) the (C$_{44}$N$_2$H$_{22}$)$^+$ intensity. Reprinted from [33] with permission, copyright 2015, Elsevier.

and by calculating Ξ, the ability to select secondary ion signals that have the least matrix effect. This is the first study that has closely looked at the matrix effects occurring during the depth profiling of organic materials with the argon cluster ion beam, however, further work is needed as only a very small set of sputtering parameters have been examined, and a limited set of samples.

References

[1] Clegg J B 1991 *Growth and Characterisation of Semiconductors* ed R A Stradling and P C Klipstein (Bristol: Adam Hilger)

[2] Benninghoven A, Rüdenauer F G and Werner H W 1987 *Secondary Ion Mass Spectrometry—Basic Concepts, Instrumental Aspects, Application and Trends* (New York: Wiley)

[3] Ziegler J F Biersack TRIM

[4] Winters H F and Coburn J W 1976 *Appl. Phys. Lett.* **28** 176

[5] Bolbach G, Viari A, Galera R, Brunot A and Blais J C 1992 *Int. J. Mass Spectrom. Ion Process.* **112** 9

[6] Wilson R G, Stevie F A and Magee C W 1989 *Secondary Ion Mass Spectrometry: a Practical Handbook for Depth Profiling and Bulk Impurity Analysis* (New York: Wiley)

[7] Liau Z L, Tsaur B Y and Mayer J W 1979 *J. Vac. Sci. Tech.* **16** 121

[8] Andersen H H 1979 *Appl. Phys.* **18** 131

[9] Littmark U and Hofer W O 1980 *Nucl. Instrum. Methods* **168** 329

[10] Williams P 1980 *Appl. Phys. Lett.* **36** 758

[11] Williams P and Baker J E 1981 *Nucl. Instrum. Methods* **182/183** 15

[12] Liu R, Ng C M and Wee A T S 2003 *Appl. Surf. Sci.* **203–204** 256–9

[13] Wehner G K and Hajicek D J 1971 *J. Appl. Phys.* **42** 1145

[14] Hofer W O and Liebl H 1975 *Appl. Phys.* **8** 359

[15] Tompkins H G 1987 *Surf. Int. Anal.* **10** 105

[16] McPhail D S, Dowsett M G, Fox H, Houghton R, Leong W Y, Parker E H C and Patel G K 1988 *Surf. Int. Anal.* **11** 80

[17] Deline V R, Reuter W and Kelley R 1986 *SIMS V* ed A Benninghoven, R J Colton, D S Simons and H W Werner (Berlin: Springer) p 299

[18] Weast R C 1987 *Handbook of Chemistry and Physics* 67th edn (Boca Raton, FL: CRC)

[19] Hues S M and Williams P 1986 *Nucl. Instrum. Methods Phys. Res.* B **15** 206

[20] Homma Y and Wittmaack K 1990 *Appl. Phys.* A **50** 417

[21] Boudewijn P R and Vriezema C J 1988 *SIMS VI* ed A Benninghoven, A M Huber and H W Werner (New York: Wiley) p 499

[22] Vriezema C J, Janssen K T F and Boudewijn P R 1989 *Appl. Phys. Lett.* **54** 198

[23] Maier M, Bimberg D, Baumgart H and Phillip F 1982 *SIMS III* ed A Benninghoven, J Giber, J Laszlo, M Riedel and H W Werner (Berlin: Springer) p 336

[24] Vandervost W, Remmerie J, Shepherd F R and Swanson M L 1986 *SIMS V* ed A Benninghoven, R J Colton, D S Simons and H W Werner (Berlin: Springer) p 288

[25] Kilner J A, McPhail D S and Littlewood S D 1992 *Nucl. Instrum. Methods Phys. Res.* B **64** 632

[26] Sze S M 1985 *Semiconductor Devices, Physics and Technology* (New York: Wiley) p 451

[27] Vandervost W, Shepherd F R, Swanson M L, Plattner H H, Westcott O M and Mitchell I V 1986 *Nucl. Instrum. Methods Phys. Res.* B **15** 201

[28] Hösler W and Palmer W 1993 *Surf. Int. Anal.* **20** 609

[29] Mahoney C M 2010 *Mass Spectrom. Rev.* **29** 247–93
[30] Havelund R, Licciardello A, Bailey J, Tuccitto N, Sapuppo D, Gilmore I S, Sharp J S, Lee J L S, Mouhib T and Delcorte A 2013 *Anal. Chem.* **85** 5064–5070
[31] Mollers R, Tuccitto N, Torrisi V, Niehuis E and Licciardello A 2006 *Appl. Surf. Sci.* **252** 6509–12
[32] Czerwinski B and Delcorte A 2013 *J. Phys. Chem.* C **117** 3595–604
[33] Shard A G, Spencer S J, Smith S A, Havelund R and Gilmore I S *Int. J. Mass Spectrom.* **377** 599–609

An Introduction to Time-of-Flight Secondary Ion
Mass Spectrometry (ToF-SIMS) and its Application
to Materials Science

Sarah Fearn

Chapter 5

Application to materials science

Secondary ion mass spectrometry is an incredibly versatile characterisation technique. So long as a sample can be put into a UHV chamber without out gassing, it is possible to analyse it. Insulating samples may be interrogated with adequate charge compensation, and large sample size can also be accommodated depending on the individual instrumentation. A consequence of this versatility means that no review of the application of SIMS can be exhaustive. In the following section a range of materials and notable applications have been reviewed, which will hopefully display the diversity and range of areas where ToF-SIMS has been successfully used.

5.1 Biomaterials and tissue studies

Since the mid-1980s secondary ion mass spectrometers have been successfully applied to the ion mapping of tooth sections, ion distribution studies of $^{44}Ca^+$ isotope exchange experiments in bone [1, 2] as well as elemental distributions between bone and Ti implants [3]. Bushinsky *et al* [4–9] continued through the 1990s extensively using secondary ion mass spectroscopy to study the relative ratios of Na^+ and K^+ ions with respect to Ca^+ and the corresponding distribution of the Na^+ and K^+ ions relative to the bone mineral and the organic bone phase during metabolic acidosis. Results showed that the organic bone phase contained the majority of the Na^+ and K^+ ions, and that during metabolic acidosis, therefore, the organic material was responsible for the buffering of excess H^+ ions, and not the bone mineral [7, 10].

These early studies exploited the strengths of the SIMS instrumentation available at the time and focussed mainly on the elemental distributions of the materials being studied. The use of novel primary ion sources has now enabled more detailed information pertaining to bone mineralisation in tissue and implants to be obtained [11, 12]. With these ion sources, ToF-SIMS has the unique ability to analyse both the

doi:10.1088/978-1-6817-4088-1ch5

mineral and the organic phases of bone tissue simultaneously, lending itself perfectly to studies of bone-implant interfaces and a range of mineralisation diseases [12].

More recently, research into bone regeneration using silicon-based scaffolds has been done using ToF-SIMS [13]. For the optimisation of such silica-based bone grafts, an important step is calcium incorporation. The calcium is vital for the formation of the hydroxycarbonate apatite (HCA) layer, and has a great influence on stimulating new bone formation. Therefore, key to the development and improvement of such biomaterials will be an understanding of the distribution of the critical elements (silicon, which is the backbone of the inorganic network, and calcium) in the 3D network.

These hybrid biomaterials of bone scaffolds pose difficulties for ToF-SIMS analysis, as they are constructed from a network of interconnected pores. Two methodologies were therefore assessed by Wang et al in order to achieve the most useful SIMS data. Initially, a 'soak and solid' approach was tested, whereby a resin is used to fill the pores and form a flat sample surface. However, for the inorganic/organic hybrid scaffolds the 'soak and solid' technique was found to mask compounds of interest. The second preparation technique was to carry out a 'mark and map' process where fiducial marks are made in the regions of interest on scaffold struts using a focussed ion beam (FIB). Using this procedure it was these regions of interest that were easily identified in the ToF-SIMS, and Wang was able to ion map the distribution of Si, Ca and other small molecules as shown in figure 5.1. The maps show an uneven distribution of Si and Ca over the scaffold strut region, indicating that the calcium has not been well incorporated into the structure. The results indicated that the use of soluble calcium phosphate

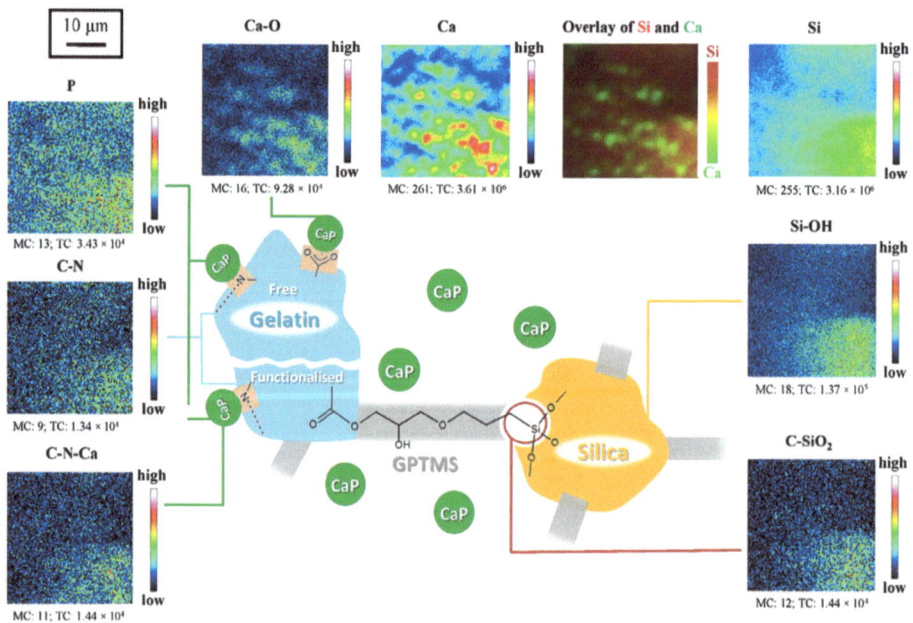

Figure 5.1. Schematic of a calcium containing silica/gelatin hybrid scaffold and ToF-SIMS secondary ion maps for critical elements and compounds. Reprinted from [13] with permission, copyright 2014, IOP Publishing.

was not an optimal precursor for the sol-gel hybrid synthesis, and other routes for effective calcium incorporation needed to be identified.

Brunelle and co-workers have been prolific in applying ToF-SIMS imaging to a range of tissue studies. Initially exploiting the Au_3^+ ion source to study lipid distributions in mouse tissue with Duchene muscular dystrophy [14], and then using the Bi_3^+ LMIG to image lipid biomarkers of Fabry disease [15], human non-alcoholic fatty liver disease [16] and more recently fatty acids in human skin [17], to understand the lipid distribution and therefore the passive diffusion of drug molecules in through the skin. The common theme to all of the studies mentioned is that they are all focussed on the localisation and imaging of lipids. Complex lipids such as phosphatidylcholine, phosphatidylethanolamines, sulfatides and glycosphin-golipids can be detected, along with low mass compounds such as fatty acids, cholesterol, cholesterol sulphate, bile acids and vitamin E can all be imaged [18].

Focussing on one important lipid—cholesterol, it is believed that this lipid is elevated in Alzheimer's disease (AD) sufferers. Understanding the distribution of this particular lipid in brain tissue could, therefore, lead to further insight into the progression of this disease. Most studies concerning brain analyses have been performed on either rat or mouse tissue [19, 20]. In a study carried out by Lazar et al [21] ToF-SIMS ion imaging was carried out on human cortex tissue to map the localisation of cholesterol in the cortical layers, figure 5.2. A series of ion images were taken from the pial (outside) surface to the white matter of the tissue, see figure 5.2(b), and the cholesterol count from each image then plotted as a function of distance from the outer pial surface, figure 5.2(c). Analyses were carried out on five control 'healthy' samples and six AD samples. It was observed that the cholesterol content in the grey matter of the AD tissue samples was 34% higher compared to the healthy tissues samples. These results show that without the need for staining or labelling of the tissue, ToF-SIMS ion imaging provided an accurate technique for mapping cholesterol over the tissues, clearly showing significant differences in cholesterol content between healthy and diseased tissue with respect to AD.

Historically, depth profiling of organic/biological materials had been considered impossible. The ion beam damage caused by mono-atomic ion beams meant that molecular ion signals could not be monitored. It is here that the cluster ion beam sources have created a lot of interest. On the back of earlier work showing successful depth profiles through organic films [22], Jones et al [23] used a C_{60}^+ primary ion beam to monitor lipid signals as a function depth through rat brain tissue. After careful sample preparation to remove excess native salts, they were able to depth profile from the surface of the tissue to the substrate ($\sim$4 μm). It was also observed that in order to measure stable ion signals for cholesterol it was necessary to hold the sample at −120 °C. Analyses carried out at room temperature indicated changing ion signals with respect to depth. It was suggested that at room temperature and under vacuum there is molecular migration within the tissue sample [24]. More recently, the capabilities of the Ar_n^+ cluster primary ion beam for depth profiling through a biological sample have been investigated [25]. In the Ar^+ cluster study, a selection of lipids signals were monitored as a rat brain tissue was sputtered with the Ar_n^+ sputter ion beam, and each exposed sputtered surface analysed with the Bi_3^+ beam.

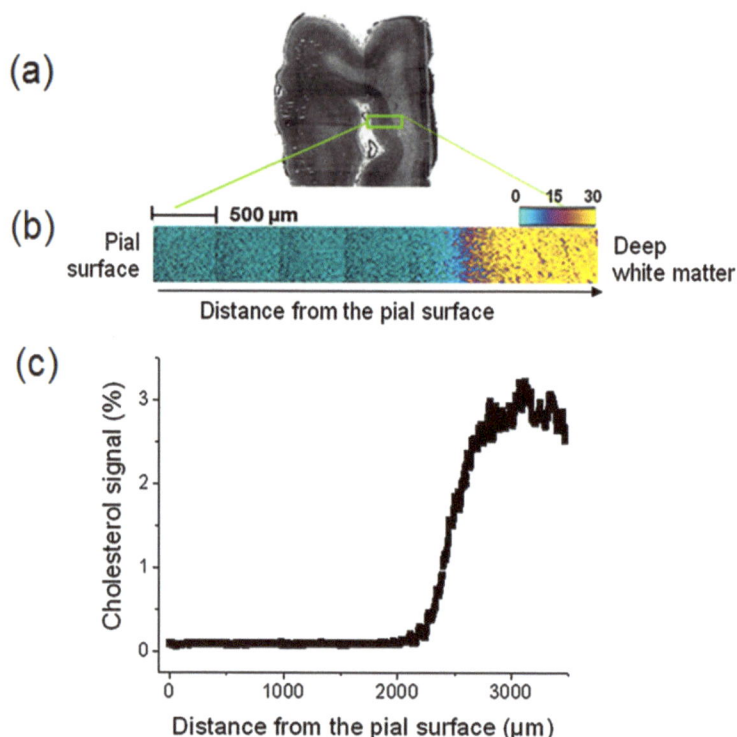

Figure 5.2. (*a*) The cryostat section of the temporal cortex (superior temporal gyrus) used for ToF-SIMS ion mapping, the selected region is highlighted in green. (*b*) The distribution of cholesterol ions mapped along the selected region from the pial to the white matter. (*c*) The cholesterol signal expressed as a percentage of the total ion intensity as a function of the distance from the pial surface. Reprinted from [21] with permission, copyright 2013, Springer Science and Business Media.

The data showed that all the signals remained constant throughout the sample thickness, figure 5.3, except for the cholesterol ion as previously reported by Jones [23]. The cholesterol ions at *m/z* 369 and 385 decrease strongly at the start of the profile, plateau out, and then decrease again as the substrate is approached. Further investigation of the cholesterol secondary ion signal showed that it is located toward the top of the tissue section. This was observed by reconstructing the sputtered volume in three dimensions, figure 5.3(*b*). The reconstruction shows the cholesterol (green) is unevenly distributed throughout the analytical volume, in particular in comparison to the phosphocholine, PC, signal (blue).

The concentration at or near the tissue surface does not appear to be an accurate distribution of cholesterol, but related to various experimental conditions. Firstly, as previously shown, sample temperature plays a role in the migration of the cholesterol ions; by reducing the sample temperature less migration is observed. Secondly, the vacuum also appears to affect the distribution of the cholesterol molecule, however, no mechanism for this phenomenon has been put forward to explain this [24]. Finally, the presence of cholesterol effects the ionisation of other molecular signals [25]. Depth profiles in standards made to assess the matrix effect, i.e. the enhancement/suppression

(a)

remaining sample thickness (%)

(b)

- cholesterol
- PC
- silicon

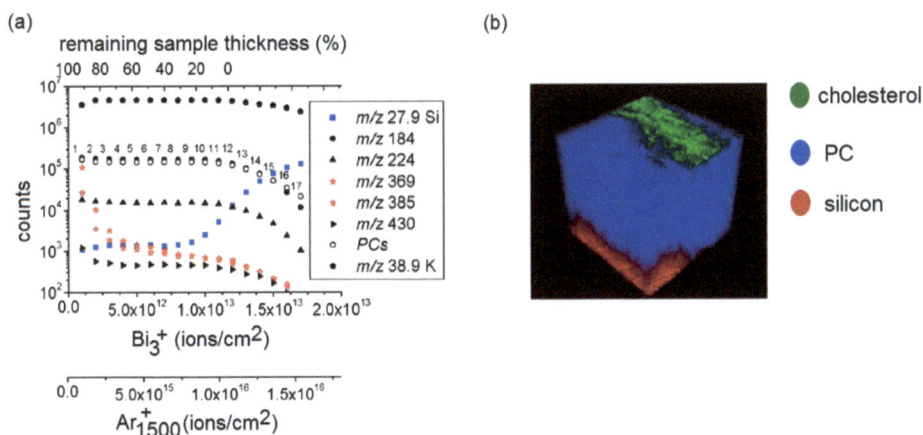

Figure 5.3. (*a*) Positive secondary ion intensities for the depth profiled rat brain tissue using an Ar_n^+ sputter ion beam. (*b*) A 3D reconstruction of the sputtered volume. Reprinted (adapted) from [25] with permission, copyright 2013, American Chemical Society.

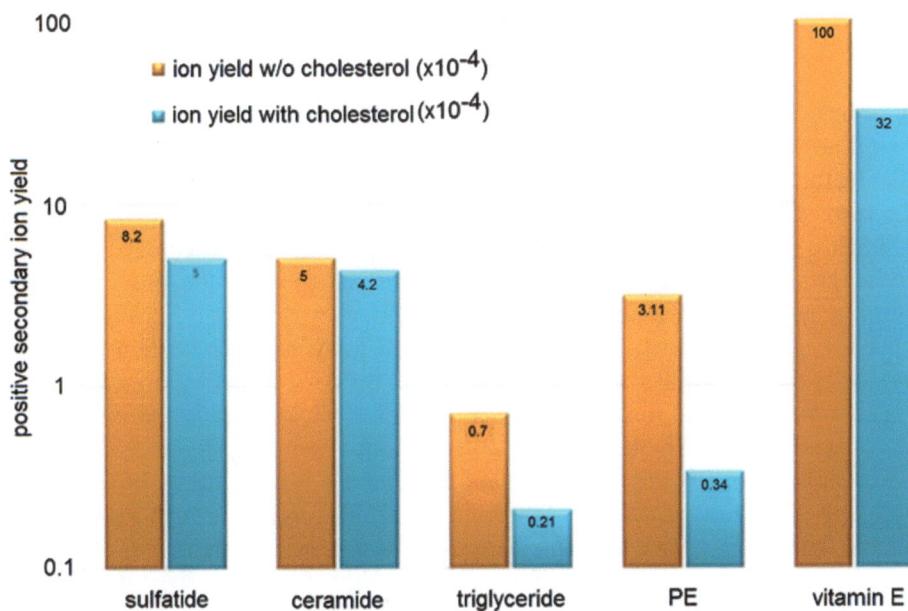

Figure 5.4. Positive secondary ions of selected lipids sputtered, with (w), and without (w/o) the presence of cholesterol. Reprinted (adapted) from [25] with permission, copyright 2013, American Chemical Society.

of ion species due to its local chemical environment, clearly showed that the secondary ion signal of other lipids is suppressed. The positive secondary ions of the selected lipids, with, and without the presence of cholesterol, are plotted in figure 5.4. The suppression of lipid ion signals and enhancement of cholesterol will, therefore, result in depth profiles that misrepresent the molecular ion distribution through tissue samples. As the work by Bich *et al* shows, fundamental studies of the ion beam interactions in biological

substate(blue)

phospholipids (red)

aminoacids (green)

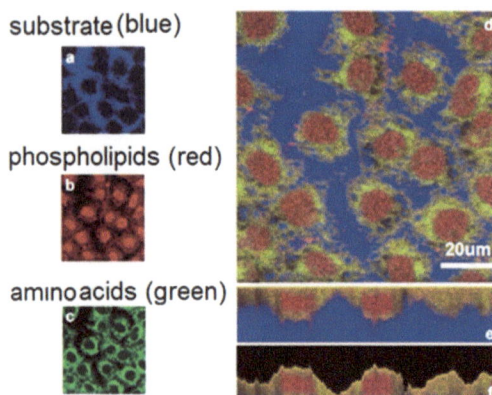

Figure 5.5. Red–green–blue colour overlay for correlation analysis of the data. Pooled signals of amino acid fragment ions are represented in red (*b*), those of phospholipids in green (*c*), and substrate-derived secondary ions are depicted in blue (*a*). Horizontal *x–y* sections are shown in (*a*)–(*d*), and vertical *x–z* sections through the sample are shown in (*e*), and a sputter corrected image in (*f*).

tissue are essential. Without understanding these phenomena, it can be very easy to misinterpret SIMS data.

With the advances in ToF-SIMS instrumentation comes the drive to test the limitations of the technology and increase our expectations of what can be achieved. Once it is possible to extract as much chemistry as possible whilst imaging smaller and smaller features we want to do this on as many dimensions as possible. As shown earlier in the reconstruction of the SIMS data in three dimensions of the rat brain tissue (figure 5.3(*b*)), attempts at the 3D visualisation of cells have also been carried out by two other groups [26, 27]. In both cases the approach was to sputter the cells with a C_{60}^{+} sputter ion beam, and then analyse the exposed surface with a LMIG primary beam. The data was then reconstructed post ToF-SIMS imaging. Breisentstein *et al* [26] analysed layers of normal rat kidney cells grown on cover slips. Figures 5.5(*a*), (*b*) and (*c*) show the totalled ion signal for the substrate, phospholipids and amino acid fragments respectively. An overlay of these three groups is shown in figure 5.5(*d*) clearly showing the cells against the substrate, with the amino acids indicating the nucleus of the cell, and the lipid signal the cell membrane. The depth profile through the sample is shown in figure 5.5(*e*), and the data was then reconstructed to correct for uneven sputtering through the sample figure 5.5(*f*). The resolution of the analysis was 350 nm laterally, making it possible to determine chemical composition on a sub-cellular level, while in the vertical dimension, resolution was estimated to be ~100 nm.

Sample preparation

In order to benefit from the high surface sensitivity possible with ToF-SIMS analyses, sample preparation is key to obtaining high quality data. Two basic requirements of the samples to be analysed are: (i) the sample is vacuum compatible as the analysis occurs in ultra-high vacuum conditions (10^{-10} mbar) and (ii) that the sample surface presented to the analytical ion beam is flat, as surface roughness will reduce the mass resolution of the analysis. In the case of biological samples, it is also essential that the biological integrity of the samples is maintained.

For the ToF-SIMS analysis of biological tissue a relatively simple and reproducible sample preparation has been routinely used by Brunelle and co-workers [28]. The biological material is first frozen in liquid nitrogen or isopentane, then sliced at $-18/20\,°C$ on a cryostat producing tissue slices ~ 15 µm thick. These slices are then deposited onto either stainless steel or glass plates and kept at $-80\,°C$. Prior to SIMS analysis the plates are warmed to room temperature and dried under vacuum. This sample preparation has been shown to maintain the chemistry of the sample and no delocalisation of compounds has been observed, including lighter elements such as Na^+ and K^+ ions. Furthermore, the samples may be used in other imaging techniques post ToF-SIMS analysis.

The analysis of biological cells requires another sample preparation procedure. Malm *et al* [29] investigated the effect of cryofixation (after washing with ammonium formate and freeze drying) and glutaraldehyde (GA) fixation and drying, on the morphology and chemical structure of human fibroblast cells (hTERT). The lipid distribution and large-scale morphology of the cells were found to be retained by both methods. However, important differences between the two preparation techniques were observed: the cryofixed cells were found to have more intact cell membranes, retained ion distribution, and gave higher ion yields for molecular phosphatidycholine (PC) ions. In contrast the GA chemical fixation with alcohol drying produced samples that had more cell structure but did not retain ion distribution. Therefore, depending on the information required judicious sample preparation must be selected. More complicated sample procedures such as freeze fracturing have also been suggested for the analysis of cells [30]. Initially this procedure was deemed unreproducible, but the development of specialist sample stages has improved the results possible via this procedure [31].

Surface contamination from polydimethylsiloxane (PDMS) can be a major problem with regard to biological samples, in particular ones that have been embedded in resin. This was observed during a recent study to identify silicates in mineralised nodules formed from osteoblasts [32]. The Quetol embedded samples were ultra-microtomed with a diamond knife. Mass spectra from the surface of the sample showed the characteristic peaks of PDMS. The surface was sputter cleaned using a C_{60}^+ ion beam and ion images were taken after the sputter cleaning. Figure 5.6 shows the extent of the surface contamination. On the left hand side of both ion images is the un-sputtered region, which is covered in silicon as shown by the Si^+ ion image, figure 5.6(*a*). On the right hand side of the ion images the PDMS has been removed, and the structure of the collagen is now clearly visible on figure 5.6(*b*). The Ca^+ ion image also shows how the presence of PDMS on the samples surface suppresses the Ca^+ ion signal. Once removed the distribution of the Ca^+ ion signal is more clearly visible.

Sectioned biological tissue may also contain high levels of Na^+ and K^+ ions. These native salts must be removed before SIMS analysis as they can lead to unreliable secondary ion signals. Washing the samples for 30 s in a 0.15 M ammonium formate solution has been shown to successfully remove the excess salts [33].

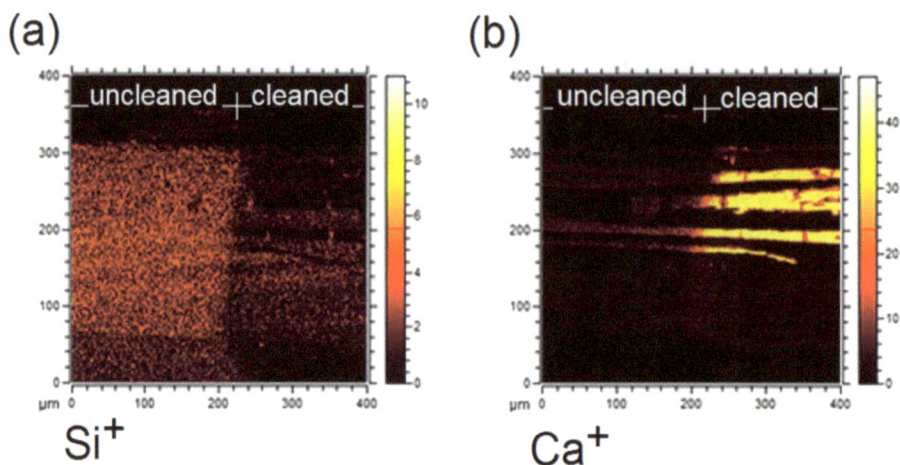

Figure 5.6. Surface contamination on a microtomed surface of the Quetol embedded mineralised nodules. (*a*) The Si^+ and (*b*) the Ca^+ ion image showing the 'as received surface' un-sputtered surface on the left hand side of the ion images and the surface after sputtering with a C_{60}^+ ion gun on the right hand side. Reprinted from [32] with permission, copyright 2013, The Royal Society of Chemistry.

5.2 Glass corrosion

Conventional depth profiling of glasses has been carried out on glass materials for many years. It has been used to measure changing compositions in order to identify problems due to manufacturing, the effects of processing, and storage. SIMS depth profiling has also been applied to many investigations of glass corrosion, for example in the study of museum glass [34, 35] and particularly in the field of nuclear waste management [36, 37]. Future strategies for the immobilisation of nuclear high level waste (HLW) are to vitrify it and dispose of it below ground up to depths of 1000 m. Dissolution studies of nuclear HLW glass are, therefore, of great interest due to the potential long-term effects of HLW on the environment. Glass instability and leaching may arise, for example, due to interactions with underground water [38]. Previously, SIMS has been used in several studies to examine the elemental distribution in corroded HLW glass SON68 and SM513 after long-term leaching experiments [39, 40].

Heavily corroded HLW glasses, however, are not straightforward samples to depth profile using a dual-beam ToF-SIMS instrument. Difficulties arise for several reasons: the surfaces of the corroded samples become extremely rough due to the exchange of material during the leaching experiments, so the depth resolution of the depth profiles will be very poor. Charge compensation is also necessary as the samples are insulating. In a dual-beam ToF-SIMS, during depth profiling the primary ion beam first scans the selected area, and the sputter ion gun is used to remove material, thereby exposing the next layer of sample to be analysed by the primary ion beam. At the start of the analysis it may be possible to charge compensate the very low current of the primary ion beam at the surface, but the current of the sputter gun is much, much higher. As the samples are highly insulating the positive charge from the sputter ion gun builds up on the surface. If the energy of the electron flood gun is low it is not possible to charge compensate the samples sufficiently to enable the sputtering and analysis of the sample. The secondary ion signal drops off very quickly, and no more ions are sputtered and detected. A novel way of analysing the sample is needed.

One way of overcoming the difficulties of the surface roughness of the sample and poor to no charge compensation, is to remove the need to depth profile into the samples. This can be achieved by fabricating a bevel or slope into the sample. The composition and buried features of the sample are then exposed as magnified surface features. Bevels or slopes can be successfully created on the glass samples using a FIB instrument. The resulting ion beam milled surface is also very smooth compared to the top of corroded glass surface. Once the bevel has been successfully fabricated the sample is transferred to the ToF-SIMS instrument for ion mapping. The location of the bevel is made by using fiducial marks that are also made during the ion beam milling process, which can be easily identified in the ToF-SIMS instrument. Once in the ToF-SIMS, instead of needing to depth profile into the sample, ToF-SIMS ion mapping can be carried out using the much lower primary ion beam currents, which are much easier to charge compensate. An example of elemental ion maps obtained from bevels made into a HLW glass leached for 7 d are shown in figure 5.7 [41]. The roughness and cracking of the surface of the samples can be seen very clearly in the Na^+ and Li^+ ion images, and the distribution of the different elements within the glass is clearly visible.

The SIMS ion images shown in figure 5.7 can be easily converted into depth profiles. Each of the individual ion images is made up of a matrix of pixels 256×256, with each pixel corresponding to a specific ion count for the element selected, figure 5.8(b). This matrix of ion count data can be integrated over its width to form a single column of ion count data. By repeating the integration step for the entire set of ion images taken for each of the elements, they can then be plotted as depth profiles once the depth scale is calculated.

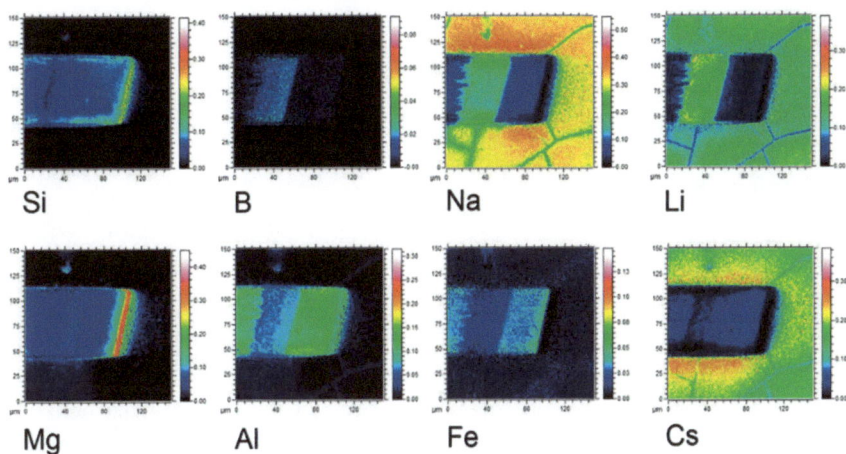

Figure 5.7. Normalised positive ion images of seven day leached glass. Reprinted from [41] with permission, copyright 2014, Elsevier.

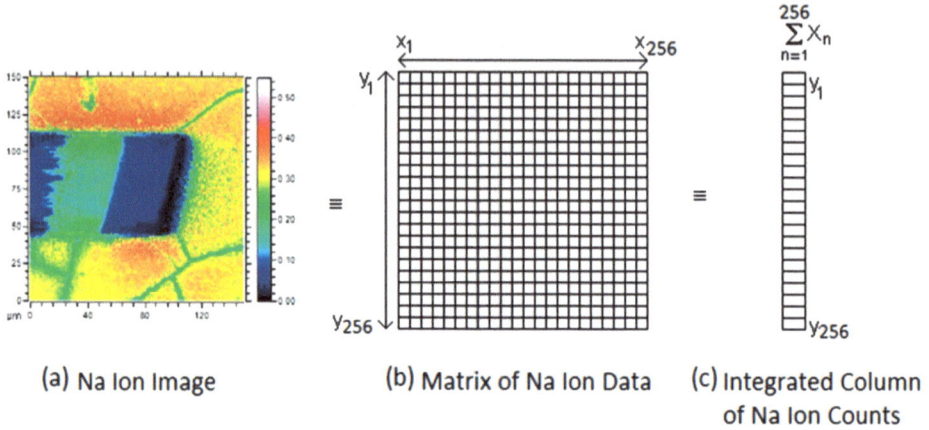

(a) Na Ion Image   (b) Matrix of Na Ion Data   (c) Integrated Column of Na Ion Counts

Figure 5.8. The ion map-linescan data conversion. (*a*) The individual ion images are made up of a matrix of data (*b*) 256×256 pixels. This matrix may be integrated over its width (*c*) to form a single column of boron count data.

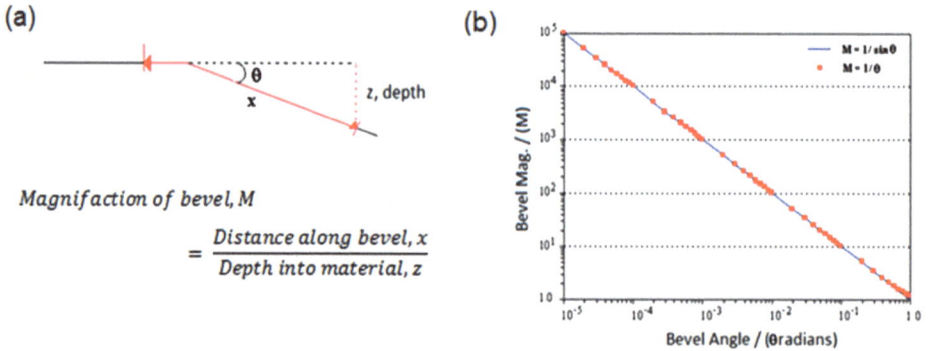

Figure 5.9. (*a*) The relationship between the bevel magnification, *M*, and bevel angle, θ, needed for the conversion of the ion image distance, *x*, along the bevel to depth *z*. (*b*) The comparison between $M = 1/\sin\theta$ and $M = 1/\theta$ for angles between 10^{-1} and 10^{-6} rads.

The depth scale is obtained using simple trigonometry, figure 5.9(*a*), and the depth scale is calculated using the magnification, *M* of the bevel which is calculated using the equation below [42]:

$$\text{Magnification of bevel, } M = \frac{\text{distance along bevel, } x}{\text{depth into material, } z}$$

$$M = \frac{x/(\sin\theta)}{x} \qquad (5.1)$$

where, *x*, is the length of the ion image made in the ToF-SIMS (the field-of-view (FoV)) and θ the angle of the bevel. This must be accurately measured using an interferometer or stylus profilometer. From equation (5.1) it can be seen that $M = \frac{1}{\sin\theta}$ for θ in radians and when θ is small, *M* is nearly the same as the

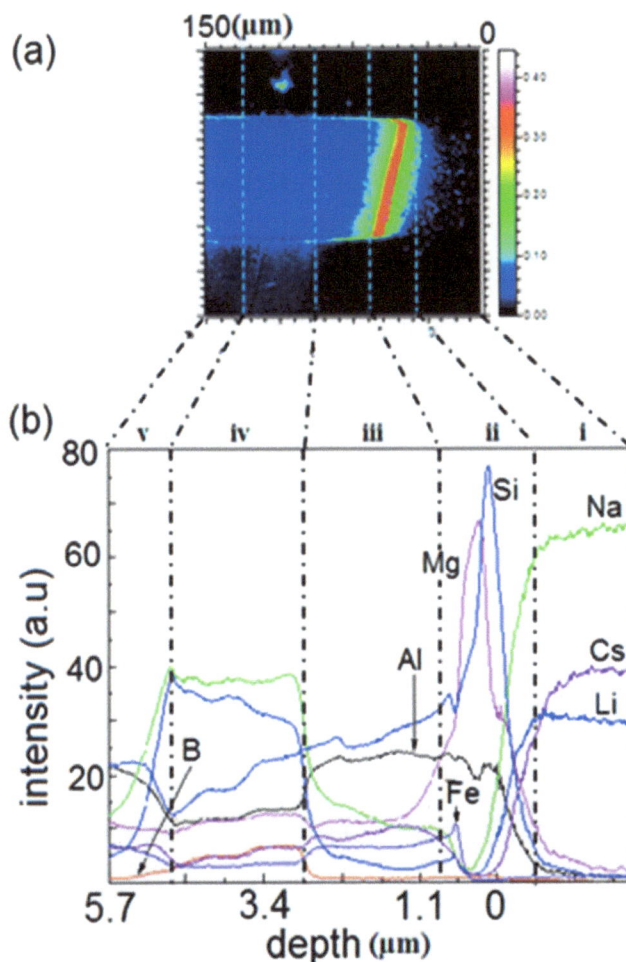

Figure 5.10. The depth profile obtained by doing an ion map-linescan ToF-SIMS analysis of a 5° bevel made into a HLW glass via FIB milling. The length of the ion image over the bevel is converted into depth. Reprinted from [41] with permission, copyright 2014, Elsevier.

reciprocal of the bevel angle so, $M \approx \frac{1}{\theta}$. Figure 5.9(*b*) below shows the comparison between $M = \frac{1}{\sin\theta}$, and $M = \frac{1}{\theta}$ for angles between 10^{-1} and 10^{-6} rads. The depth into the sample, z, therefore, can be calculated as being (the distance along bevel, x)/ (bevel magnification, M).

Once the FoV of the SIMS ion image has been converted into the corresponding depth scale, a depth profile can be plotted as shown in figure 5.10. The 'FoV at the x-axis', has been converted into depth and the different regions of interest are also marked out in the figure to show more clearly the changing chemical composition in the glass due to the leaching and diffusion of the elements in the glass. A further advantage of this technique is that a much greater depth into the sample can be reached. The final depth scale of the depth profile shown in figure 5.10 is 5.70 µm.

It would be impossible to depth profile up to such a depth in a SIMS analysis as the depth resolution would be quickly lost within the first micron due to the initial roughness of the sample surface.

This sample preparation: FIB-bevelling of the sample, although demonstrated on highly corroded nuclear waste glass, may also be useful for other samples whereby the initial surface is too rough and too insulating, or to reveal very deeply buried features.

5.3 Ceramic oxides

The measurement of oxygen diffusion in oxide materials using SIMS has significantly developed since the early experiments of the 1960s. A growing interest in oxygen ion conducting materials for solid oxide fuel cells (SOFC) and oxygen permeation membranes (OPC) has been mainly responsible for this development. The extensive use of SIMS in the field of solid-state ionics is explained by the technique's ability to distinguish between stable isotopes as a function of position in the solid sample. For a material whose device performance is dependent on the rate at which oxygen either enters into or passes through an oxide, determining the kinetics of this process is essential. SIMS enables stable isotope tracer diffusion profiles to be accurately and precisely measured, and diffusion coefficients can be obtained. Typically SIMS profiles are affected by the matrix effects of the material and limit quantification, however, as the ratio of the chemically identical isotopes is measured these effects are nullified.

The isotope tracer diffusion experiment is a relatively simple procedure. It involves initially equilibrating a bulk ceramic material in an oxygen atmosphere of natural abundance, followed by exposure to an enriched ^{18}O atmosphere at the same temperature and oxygen partial pressure, for a known period of time. After quenching the sample the ^{18}O diffusion profile is measured using SIMS (secondary ion mass spectrometry) [43, 44]. For short diffusion profiles (up to ~2 μm) depth profiling can be carried out. For diffusion profiles that extend up to several hundred microns, an ion map-linescan ToF-SIMS analysis, as described in the previous section 5.2, is carried out, on a cross-section of the sample [45].

To obtain values for the diffusion coefficient, D^* (and the surface exchange coefficient k^*), the measured profile of ^{18}O isotopic fraction $C'(O)$ is fitted to the solution of Fick's 2nd law of diffusion for a semi-infinite medium, shown below in equation (5.2) [46], using a regression analysis fit in Matlab [47].

$$C'(O) = \frac{C_x - C_{bg}}{C_g - C_{bg}} = \text{erfc}\left(\frac{x}{2\sqrt{D^*t}}\right) - \exp\left(hx + h^2D^*t\right)\text{erfc}\left(\frac{x}{2\sqrt{D^*t}} + h\sqrt{D^*t}\right)$$

(5.2)

where $h = \frac{k^*}{D^*}$, C_x is the ^{18}O isotopic fraction at depth x, C_g the isotopic fraction of ^{18}O in the gas phase and C_{bg} the ^{18}O background isotopic fraction, and t the ^{18}O anneal time. ^{18}O isotopic fraction is obtained by measuring the ^{18}O$^-$ and ^{16}O$^-$ secondary ions during the SIMS analysis.

For bulk diffusion, transport parameters over many orders of magnitude can be obtained using the tracer isotope diffusion SIMS experiment. The range of diffusion

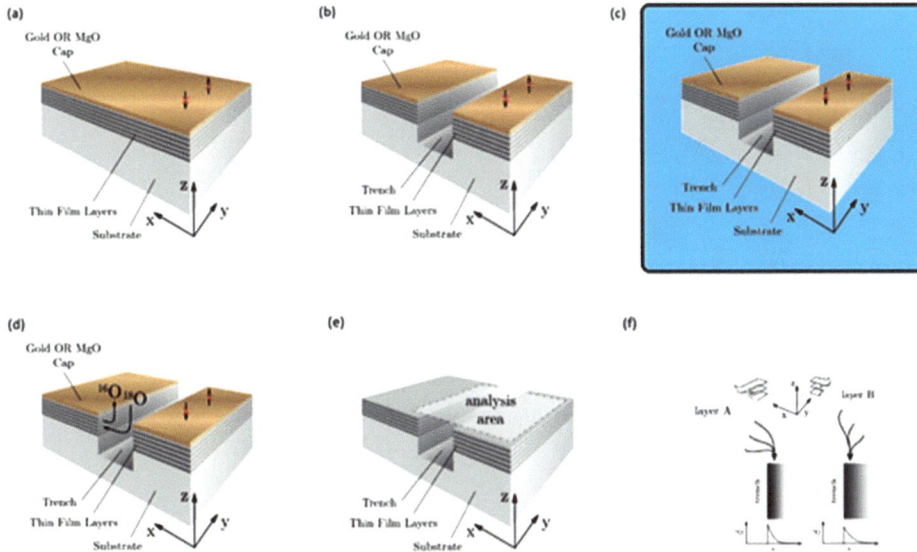

Figure 5.11. (*a*) The multi-layer sample is gold coated to form a blocking layer. (*b*) A trench is made into the sample to expose the layer edges. (*c*) The sample is exposed to the 18O enriched atmosphere and (*d*) diffusion occurs along the layers. (*e*) ToF-SIMS analysis is made over the edge of the trench along and into the sample via Cs$^+$ ion sputtering. (*f*) Ion images are extracted and integrated to form linescans for each layer. Reprinted (adapted) from [50] with permission, copyright 2012, American Chemical Society.

coefficients that can be measured via this technique can vary from 10^{-5} cm^2 s^{-1} to 10^{-19} cm^2 s^{-1} [48].

More recently, more complex diffusion problems due to the presence of grain boundaries, interfaces or strain, have been investigated with the use of pulsed laser deposited (PLD) thin film samples [49–51]. Again, these SIMS analyses are made possible due to the high spatial resolution of the 'ion map-linescan' analyses. However, before the ToF-SIMS analyses can be carried out the tracer diffusion experiment must be adapted to ensure that diffusion occurs along an axis that the diffusion can be measured. Typically this means making sure that the diffusion occurs laterally along the layers. A schematic of the adapted tracer diffusion experiment for measuring a multi-layer oxide is shown in figure 5.11. The multilayer sample is masked by a gold blocking layer, and a trench is then made into the sample, figures 5.11(*a*) and (*b*) respectively. Once exposed to the ^{18}O enriched atmosphere, figure 5.11(*c*), the gold layer will block the diffusion of the ^{18}O, and force it to diffuse along each of the layers in the structure, figure 5.11(*d*). The gold blocking layer is then removed, and SIMS ion mapping is carried out measuring the oxygen signals, figure 5.11(*e*). The sample is also sputtered away using a Cs$^+$ ion beam so that each of the layers in the structure is measured. The ion maps obtained from each layer are then converted into 2D profiles, figure 5.11(*f*).

Using this adapted tracer diffusion experiment and the ion map-linescan analyses, transport diffusion parameters were extracted from an yttria stabilised zirconia (YSZ)-ceria oxide (CeO$_2$) multilayer sample with strained interfaces. The aim of the

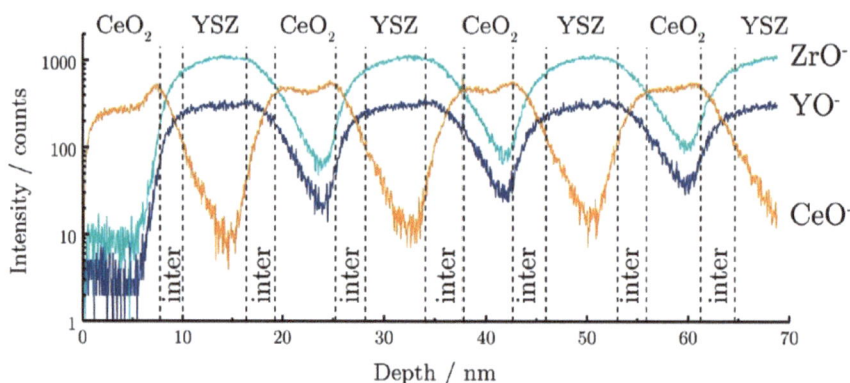

Figure 5.12. A dual-beam ToF-SIMS depth profile through the CeO_2-YSZ multilayer sample. Reprinted (adapted) from [50] with permission, copyright 2012, American Chemical Society.

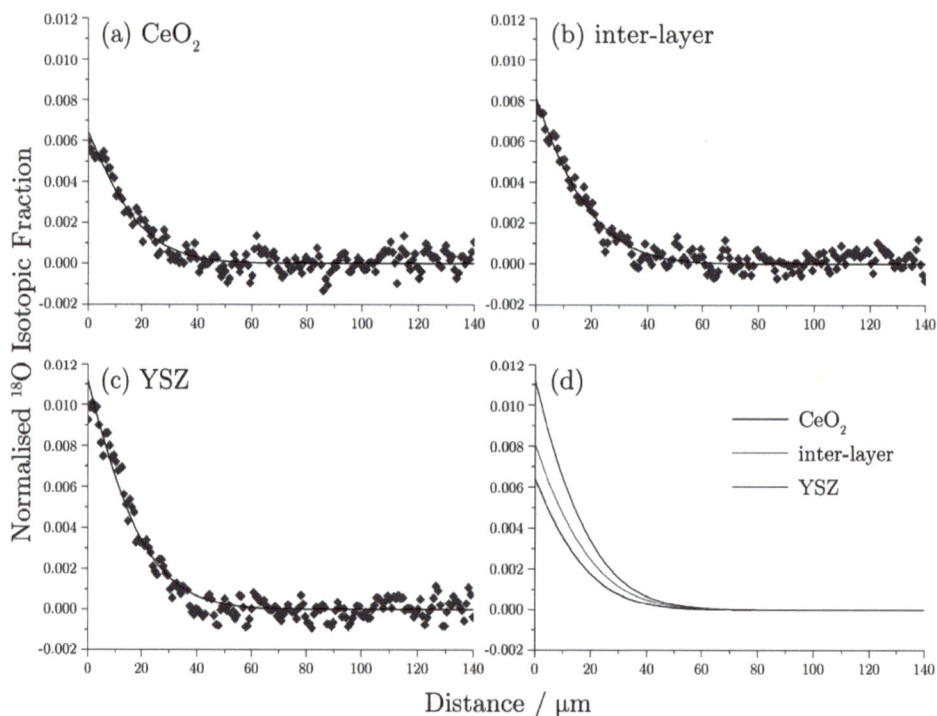

Figure 5.13. The ^{18}O isotopic fraction from (a) CeO_2, (b) inter-layer and (c) YSZ. (d) The fitted data for each layer in the structure. Reprinted (adapted) from [50] with permission, copyright 2012, American Chemical Society.

experiment was to identify any enhancement in oxygen ion transport at the hetero-interface. A SIMS depth profile of the structure is shown in figure 5.12, with the different layers clearly visible; CeO_2, YSZ and the interface (labelled 'inter'). The resulting ^{18}O diffusion linescans for each of the different regions of the structure are also shown in figure 5.13. The ToF-SIMS measurements of the oxygen isotope tracer diffusion experiments showed that the fastest oxygen diffusion was in the YSZ layers, and was comparable to the value obtained in bulk YSZ, therefore, no enhanced conductivity was observed due to the presence of the hetero-interfaces and/or the presence of strain within the PLD film.

5.4 Semiconductor analysis

Secondary ion mass spectrometry has become an everyday and powerful technique in the field of semiconductor analysis, and has been well written about in the literature over many years [52] and so only a brief description will be given of some of the areas where the technique has been used, and how the dual-beam ToF-SIMS instruments have taken the analysis from two dimensions to chemical characterisation of novel 3D architectures.

The flexibility of secondary ion mass spectrometry means that it is an excellent method for understanding many of the subtle defects that can arise in the manufacturing of device structures and wafers. For example, conventional depth profiling can be performed to identify how far an implant may have progressed into a semiconductor lattice, or how an annealing step may have changed the distribution of an implant. These are extremely important parameters to understand when fabricating new implant profiles for smaller and smaller devices. The amount of element implanted—the implant dose—is also important, therefore, accurate quantification is also required. In this area, SIMS is probably unique in being able to do this, despite the well known issues with regard to matrix effects. Quantification of implant profiles is possible as standards/reference material can be readily made. Ion implants of known dose are amongst the most commonly used type of reference material, and easily fabricated by ion implantation of the element of interest into the matrix material i.e. a 10 keV boron implant into silicon. These ion implant reference materials not only give the profile shape that is attained when they are depth profiled, but can provide information on background levels attainable during an analysis, so giving a detection limit for the implant within the selected matrix, and also the dynamic range for the measurement. They can also be used as a guide to data quality, instrumental performance and reproducibility; by carrying out regular analysis of the same reference material, using well recorded instrument conditions any inconsistences between analyses should reveal any instrumental issues. A schematic of a depth profile of an arsenic ion implant profile reference in silicon is shown in figure 5.14. Reference materials also help to reveal instrumental artefacts, in particular ion beam mixing phenomena. The reference material may stand as an ideal sample and the resulting SIMS data will indicate the deviation from the ideal profile. It is then possible to deconvolute the ideal from the real and give an estimate of the errors on the experimentally achieved profile.

The high surface mass resolution capabilities of ToF-SIMS can be superbly exploited to identify contamination that can occur on virgin silicon wafers deposited either during fabrication or during more seemingly benign situations. For example, silicon wafers are often transported from one site to another using plastic wafer boxes, which are kept extremely clean. It is extremely important to preserve the cleanliness of the wafer surface, as metal particle contamination can lead to defect formation, and

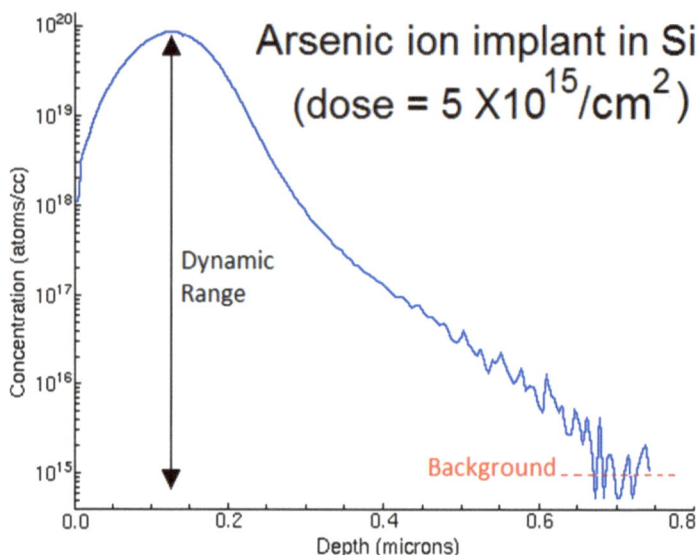

Figure 5.14. A depth profile of an arsenic ion implant profile reference, showing implant dose, dynamic range and background concentration.

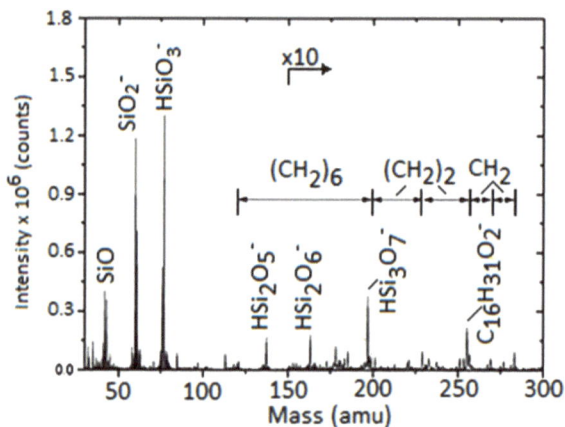

Figure 5.15. ToF-SIMS mass spectra of a fatty acid adsorbed onto a silicon wafer after storage in a wafer box. Reprinted from [53] with permission, copyright 2014, Wiley publishers.

organic residues will lead to the altering of the surface properties causing problems for subsequent film deposition or etching. A recent study highlighted the surprising deposition of fatty acids from wafer boxes onto silicon wafers [53]. The ToF-SIMS mass spectrum obtained on such a silicon wafer that had been stored in wafer box is shown in figure 5.15. The spectrum shows the Si^- containing ions, and in the high mass range the fragments from a saturated fatty acid which have the formula $C_nH_{2n-1}O_2^-$, that have adsorbed onto the wafer surface. The peak at $C_{16}H_{31}O_2^-$ is chosen as the characteristic ion for the fatty acid. The ToF-SIMS has been able to detect very clearly

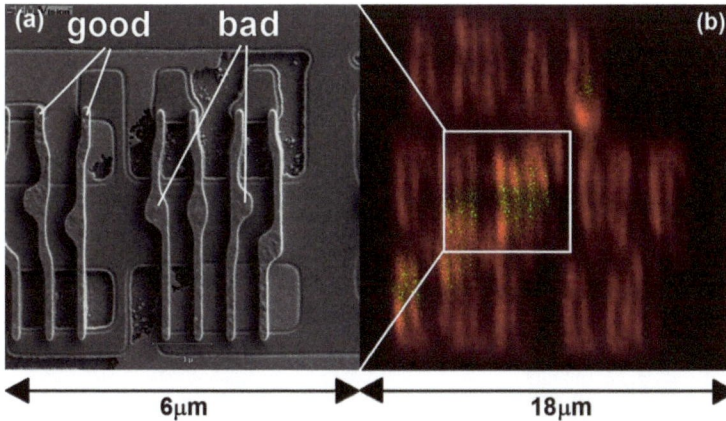

Figure 5.16. (*a*) SEM image from an area on a CMOS with defects, showing 'good' (rough) and 'bad' (smooth) polysilicon lines. These were subsequently analysed by ToF-SIMS (*b*) as indicated by the white square in the ion map overlay of mass-resolved secondary ion images. Red: Si; green: P. Reprinted from [54] with permission, copyright 2010, Wiley publishers.

the presence of the organic contamination on the surface due to the outgassing of the polymer wafer boxes. This work highlights the high surface sensitivity of ToF-SIMS.

Further to the classical SIMS analyses of depth profiling and high-resolution surface mass spectrometry, the technique is very powerful as it is also able to identify the location of defects or poor doping in devices when used to ion map species during an analysis. The example that follows shows an area of a complimentary metal-oxide semiconductor (CMOS) transistor, Figure 5.16 [54]. Phosphorous is a common species typically implanted into these devices, but must remain located in specific areas for the device to function—the so-called high voltage areas. SEM images, figure 5.16(*a*), showed that there were some 'bad' regions where the polysilicon appeared smooth, as opposed to the 'good' region where the surface was rough. The region of interest identified by the SEM images was then ion imaged using ToF-SIMS, with enough resolution to separate $30SiH^-$ and P^- secondary ion signals, but also with a high lateral resolution. The corresponding SEM image is highlighted in the white square in the ion map of figure 5.16(*b*). The overlay of figure 5.16(*b*) shows the ion maps of Si^- (red) and P^- (green). The image indicates that the regions labelled in the SEM image have elevated levels of phosphorus. The ion mapping is therefore able in this instance to locate precisely the area of device failure.

In this final example, an example is given of a 3D analysis performed on a dual-beam ToF-SIMS instrument [55]. Semiconductors such as GaAs, InGaAs and InAs have high electron mobility so are promising candidates for use in transistor devices. By integrating InGaAs channels on Si substrates high performance transistors with novel 3D architecture are being developed: trigate [56], fin shaped field effect transistors (fin-FETs) and tunnel FETs. The performance of these 3D devices will be strongly influenced by the quality and properties of the surface and interfaces of the structures. High-resolution physico-chemical characterisation is therefore needed to analyse the materials integrated into the structures.

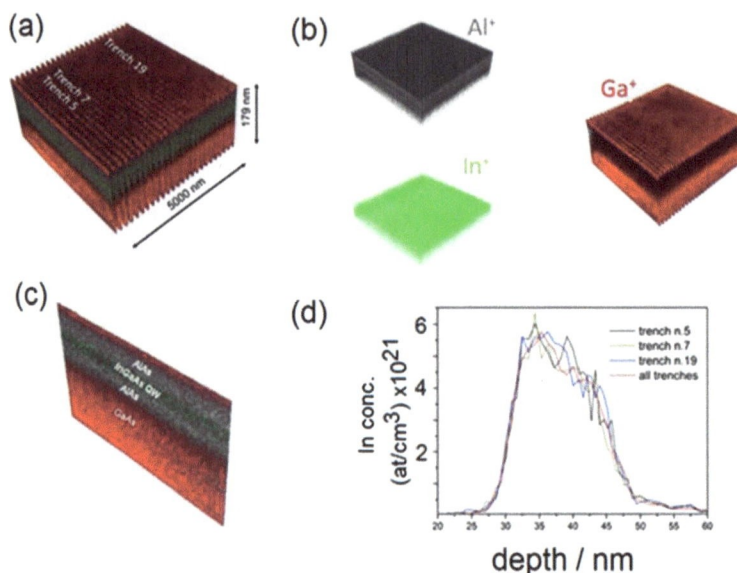

Figure 5.17. (*a*) 3D ToF-SIMS reconstruction of InGaAs QW with AlAs barrier layers, highlighting trenches 5, 7 and 19 41 nm wide. (*b*) Separate reconstruction of Al, In, Ga ion maps, (*c*) a reconstruction of an individual III–V trench of 3074 × 41 × 179 nm and (*d*) SIMS depth profiles of trenches 5, 7 and 19. Reprinted from [55] with permission, copyright 2015, Wiley publishers.

The chemical composition of InGaAs/AlAs quantum wells selectively grown in SiO_2 trenches, 100–300 nm wide, was studied using a dual-beam ToF-SIMS. A Bi_3^{2+} source was rastered over a 5 × 5 μm area with 256 × 256 pixels for analysis, an O_2^+ sputter ion beam at an energy of 500 eV was rastered over a 250 × 250 μm area for depth profiling through the structure. The ToF-SIMS measurements are shown in figure 5.17. The full 3D ion map reconstruction is shown in figure 5.17(*a*), with figure 5.17(*b*) highlighting the individual layers of the stack. One of the III-V heterostructure trenches, figure 5.17(*c*), was extracted and the In concentration was quantified with use of an InGaAs reference sample. The resulting In plot is shown in figure 5.17(*d*), which shows that the In depth profiles overlap for each trench reflecting the good uniformity of the selective III–V growth method in the structures measured.

5.5 Organic electronics

Organic electronics is a vibrant field of research comprising many different thin film technologies for application covering: printable circuitry, flexible solar cells and LEDs and sensors. One unifying factor of all such systems is the importance of the microstructure and device architecture on the performance. There is still much to be learned about the driving forces leading to the formation of microstructure and establishing relevant processing-structure-property inter-relationships [57].

Of great interest is the growing development of organic photovoltaics (OPVs). These are of interest as they offer great potential for high-throughput processing, and promise to drastically lower materials costs when compared to many of their

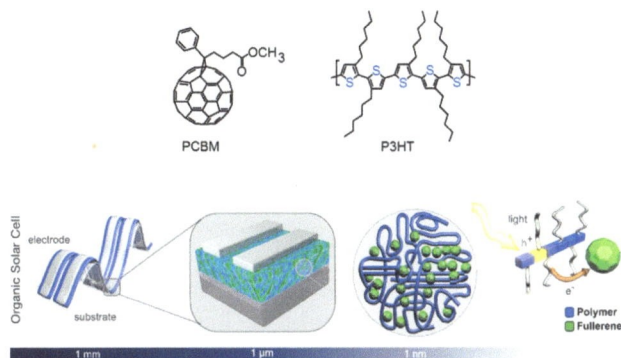

Figure 5.18. (*a*) Structure of the organic molecules used to fabricate an OPV and (*b*) the application of the organic molecules used to make an organic solar cell (figure courtesy of Dr Neil Treat).

inorganic counterparts [58]. Typically OPVs consist of a semiconducting polymer and a fullerene, such as poly(3-hexyl-thiohene)—P3HT, and [6,6]-phenyl 61-butyric acid methyl ester—PCBM, the structure of each of these molecules is shown in figure 5.18. These two materials are blended together to form a bulk heterojunction (BHJ).

The active layer of the BHJ is comprised of the electron-donating material, the P3HT, and the electron-accepting material, PCBM. Until very recently it was generally accepted that the BHJ was formed of relatively pure phases [59], however, recent studies have indicated that the BHJ microstructure may also consist of a microstructure/phase morphology that includes a finely intermixed phase [60, 61]. There still, therefore, remain fundamental questions related to the exact micro-structure of these structures. The role of and interplay between the different phases that can be present in active layers of organic solar cells, as well as their nature (e.g. size, degree of order), are believed to be critical for the resulting device performance and optimisation of their power conversion efficiencies.

One technique that holds significant promise to investigate the composition of these structures is SIMS depth profiling. Dynamic SIMS has been carried out using a single beam instrument to investigate the temperature dependant diffusion of fullerenes into different semi-conducting polymers [60]. The SIMS depth profiles highlighted that the PCBM within the BHJ active layer had significant mobility over three dimensions, even under mild heat treatments. Further studies have employed a dual-beam ToF-SIMS to investigate vertical phase separation of solution cast blends [62]. In this study a mono-atomic Cs^+ ion beam was used to perform the depth profiles, while Bi_3^+ was used to generate the secondary ions. As the use of a monoatomic ion source for sputtering results in a rapid degradation of molecular information during erosion, the ions of $^{34}S^-$, CH^- and deuterium as the PCBM were labelled (D5-PCBM).

In contrast to using mono-atomic sputtering beams, the Ar_n^+ cluster ion beam will maintain the molecular structure while eroding the samples. A dual-beam ToF-SIMS instrument could with an argon cluster sputter gun therefore accurately measure the depth and lateral distribution of the PCBM and P3HT components [63]. As molecular information is maintained, there is no need to label any of the components. Depth profiles obtained with an argon cluster ion beam, from a PCBM/P3HT organic layer on ITO, are shown in figure 5.19. The C_{60} ion signal remains constant through the sample until the ITO substrate is reached. No loss of molecular information is observed, and it can be seen that the PCBM/P3HT composition remains uniform.

As previously highlighted in figure 5.17, ion images are collected at every depth of the depth profile, so that 3D renderings of the data are possible. In figure 5.20 the negative secondary ion signals of F^-, S^-, Cl^-, C_4S^-, InO_2^- and C_{60}^- are shown as 3D ion maps. The 3D maps highlight the uniformity of the PCBM/P3HT layer throughout the structure. Interestingly the ion maps also indicate that there is an enrichment of the F^- species in the bottom right hand corner of the image. It is only by reconstructing the 3D image that the localisation of the fluorine contamination can be observed, this information would not be exposed in a conventional 2D depth profile. The enhancement of both the F^- and Cl^- ion is also visible at the ITO substrate interface.

Other successful 3D depth profiling of polymer multilayers and conductive filaments has also been carried out using the argon cluster source on a dual-beam ToF-SIMS instrument [64, 65].

As the development of organic electronics expands and the use of organic materials in everyday electronics becomes a reality, not only is a greater under-standing of their material properties needed, but also their lifetime becomes important. In this context the oxidation behaviour of another organic electronic material has been studied—rubrene [66]. Rubrene is an organic single crystal, which

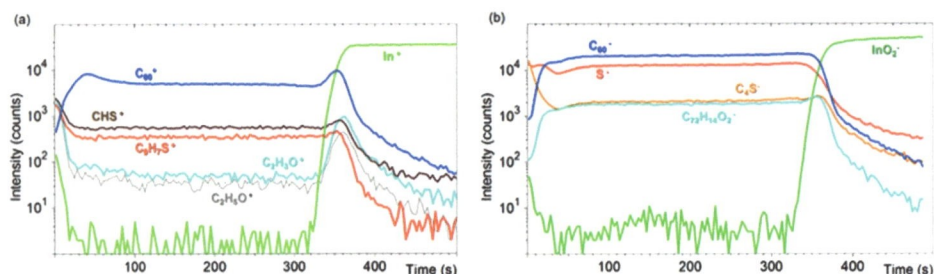

Figure 5.19. The depth (*a*) positive and (*b*) negative 5 keV Ar_{1500}^+ depth profiles from a PCBM/P3HT film on ITO. Reprinted from [63] with permission, copyright 2013, American Vacuum Society.

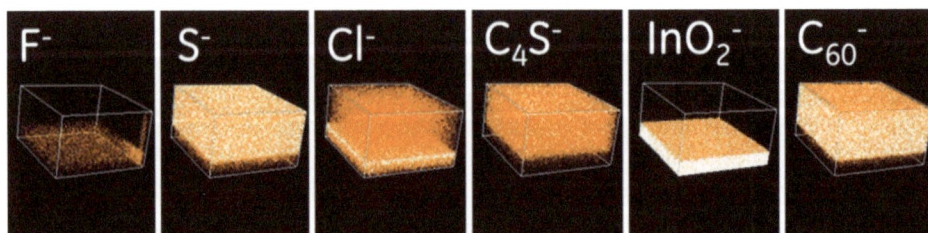

Figure 5.20. The 3D ion maps of the negative ions in the PCBM/P3HT organic layer on an ITO substrate The F and Cl are enhanced at the ITO interface. 3D image dimensions are: $x = y = 200$ μm, $z \sim 220$ nm. Reprinted from [63] with permission, copyright 2013, American Vacuum Society.

has also shown potential for application in photovoltaics due to its extremely high mobility and long exciton diffusion lengths.

The surface of rubrene (molecular formula $C_{42}H_{28}$) plays a vital role in the control of charge carrier concentration, mobility and exciton recombination. The formation, therefore, of an oxide layer on the surface will have a major effect on the material's properties and its electronic usefulness. A dual-beam ToF-SIMS was used to investigate the surface of rubrene single crystals and provide direct measurements of the molecular species present. Both the negative and positive secondary ion mass spectra are shown in figure 5.21. The parent molecule for rubrene $C_{42}H_{28}$ is clearly visible in both spectra. The rubrene isotopes are also observed and are shown in the highlight of negative ion mass spectra. Along with the rubrene signal there are several rubrene oxide signals also identified, importantly the ion $C_{42}H_{28}O_2{}^-$, rubrene peroxide.

The rubrene molecular ion, along with four oxide related ions, were ion mapped to determine their distribution over the rubrene surface, figure 5.22. The rubrene molecule ($C_{42}H_{28}$) is detected, evenly, over the ion image area, figure 5.22(a). The four oxide related ions are shown in figures 5.22(b)–(e). A native rubrene oxide ($C_{42}H_{28}O^-$) is observed to exist uniformly over the imaged area of the crystal surface. In comparison the rubrene peroxide ions ($C_{42}H_{28}O_2{}^-$) have a much more specific distribution. In figures 5.22(a) and (e) it is believed that the location of the oxide ions occurs at crystal defects. From the ion maps of figure 5.21, it can be seen that the majority of the oxygen containing ions are detected at these specific locations, whereas the rubrene signal appears depleted in the corresponding locations.

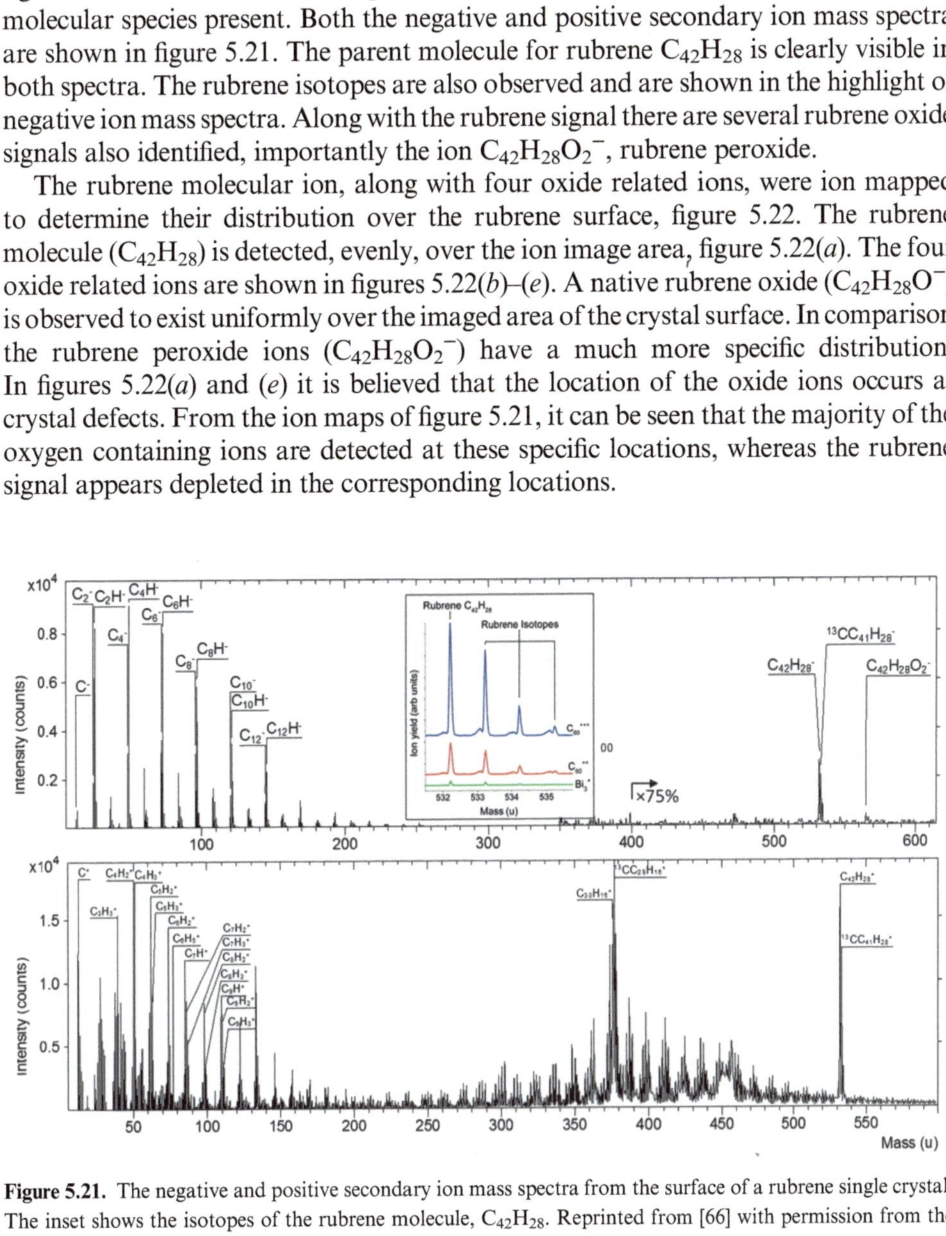

Figure 5.21. The negative and positive secondary ion mass spectra from the surface of a rubrene single crystal. The inset shows the isotopes of the rubrene molecule, $C_{42}H_{28}$. Reprinted from [66] with permission from the PCCP Owner Societies.

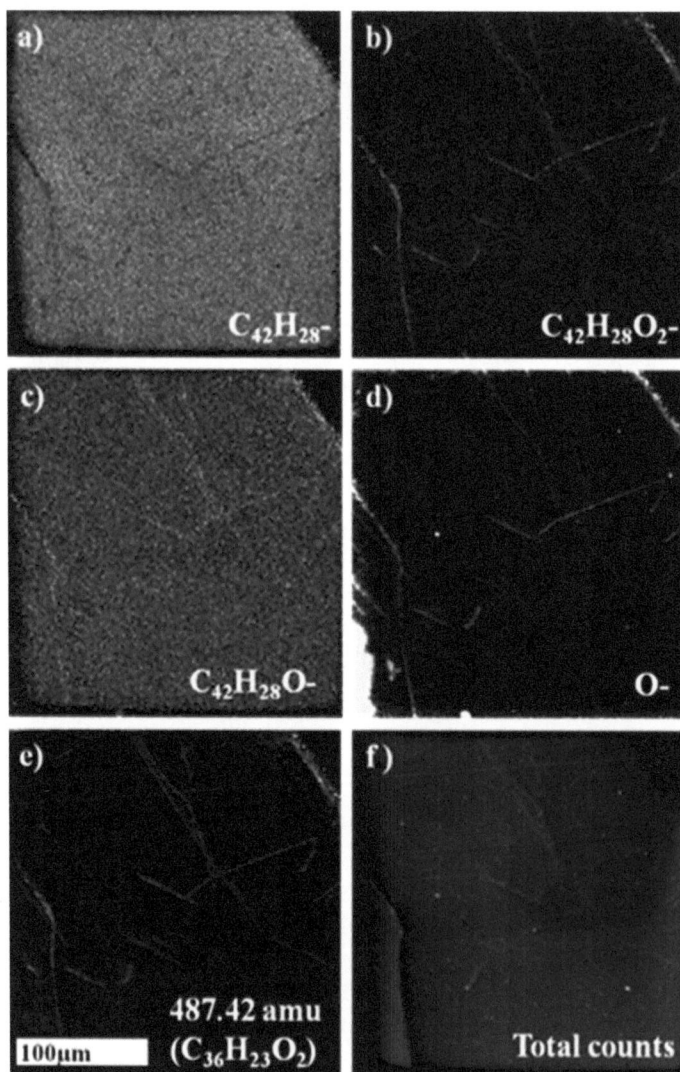

Figure 5.22. A series of normalised ion maps each showing a 250 × 250 µm area of an as grown surface of rubrene. (*a*) Rubrene, (*b*) rubrene peroxide, (*c*) rubrene oxide, (*d*) oxygen, (*e*) 487.42 amu (suggested to be $C_{36}H_{23}O_2$) and (*f*) total counts. Reprinted from [66] with permission from the PCCP Owner Societies.

This observation of local oxidation of rubrene is important as oxidation is a degradation process that can potentially ruin electronic devices. Again, although the mass spectra identified the presence of oxide species on the rubrene surface, the 2D map has been able to show the spatial distribution of these ions.

References

[1] Levi-Setti R, Wang Y L and Crow G 1986 *Appl. Surf. Sci.* **26** 249–64
[2] Bushinsky D A, Chabala J M and Levi-Setti R 1990 *Am. J. Physiol.* **259** E586–92

[3] Lodding A R, Fishcer P M, Odehus H, Noren J G, Sennerby L, Johansson C B, Chabala J M and Levi-Setti R 1990 *Anal. Chim. Acta* **241** 299–314

[4] Bushinsky D A, Chabala J M and Levi-Setti R 1989 *Am. J. Physiol.* **256** E152–8

[5] Bushinsky D A, Chabala J M and Levi-Setti R 1989 *Am. J. Physiol.* **257** E815–22

[6] Chabala J M, Levi-Setti R and Bushinsky D A 1991 *Am. J. Physiol.* **261** F76–84

[7] Bushinsky D A, Wolbach W, Sessler N E, Mogilevsky R and Levi-Setti R 1993 *J. Bone Miner. Res.* **8** 93–102

[8] Bushinsky D A, Gavrilov K, Chabala J M, Featherstone J D B and Levi-Setti R 1997 *J. Bone Miner. Res.* **12** 1664–71

[9] Bushinsky D A, Chabala J M, Gavrilov K L and Levi-Setti R 1999 *Am. J. Physiol.* **277** F813–19

[10] Bushinsky D A, Gavrilov K L, Chabala J M and Levi-Setti R 2000 *J. Bone Miner. Res.* **15** 2026–32

[11] Malmberg P, Bexell U, Eriksson C, Nygren H and Richter K 2007 *Rapid Commun. Mass Spectrom.* **21** 745–9

[12] Malmberg P and Nygren H 2008 *Proteomics* **8** 3755–62

[13] Wang D, Poologasundarampillai G, van den Bergh W, Chater R J, Kasuga T, Jones J R and McPhail D S 2014 *Biomed. Mater.* **9** 015013

[14] Touboul D, Brunelle A, Halgand F, De La Porte S and Laprevote O 2005 *J. Lipid Res.* **46** 1388–95

[15] Touboul D, Roy S, Germain D P, Chaminade P, Brunelle A and Laprevote O 2007 *Int. J. Mass Spectrom.* **260** 158–65

[16] Debois D, Bralet M-P, LeNaour F, Brunelle A and Laprevote O 2009 *Anal. Chem.* **81** 2823–31

[17] Kezutyte T, Desbenoit N, Brunelle A and Briedis V 2013 *Biointerphases* **8** 1–8

[18] Touboul D, Brunelle A and Laprevote O 2011 *Biochimie* **93** 113–19

[19] Borner K, Nygren H, Haganhoff B, Malmberg P, Tallarek E and Manson J E 2006 *Biochim. Biophys. Acta* **1761** 335–44

[20] Sjovall P, Lausmaa J and Johansson B 2004 *Anal. Chem.* **76** 4271–8

[21] Lazar A N *et al* 2013 *Acta Neuropathol.* **125** 133–44

[22] Fletcher J S, Conlan X A, Jones E A, Biddulph G, Lockyer N P and Vickerman J C 2006 *Anal. Chem.* **78** 1827–31

[23] Jones E A, Lockyer N P and Vickerman J C 2008 *Anal. Chem.* **80** 2125–32

[24] Sjovall P, Johansson B and Lausmaa 2006 *J. Appl. Surf. Sci.* **252** 6966–74

[25] Bich C, Havelund R, Moellers R, Touboul D, Kollmer F, Niehuis E, Gilmore I S and Brunelle A 2013 *Anal. Chem.* **85** 7745–52

[26] Breitenstein D, Rommel C E, Möllers R, Wegener J and Hagenhoff B 2007 *Angew. Chem.* **46** 5332–5

[27] Fletcher J S, Lockyer N P, Vaidyanathan S and Vickerman J C 2006 *Anal. Chem.* **79** 2199–206

[28] Brunelle A, Touboul D and Laprevote O 2005 *J. Mass Spectrom.* **40** 985–99

[29] Malm J, Giannaras D, Riehle M O, Gadegaard N and Sjovall P 2009 *Anal. Chem.* **81** 7197–205

[30] Roddy T P, Cannon D M Jr, Meserole C A, Winograd N and Ewing A G 2002 *Anal. Chem.* **74** 4011–19

[31] Lanekoff L, Kurczy M E, Hill R, Fletcher J S, Vickerman J C, Winograd N, Sjovall P and Ewing A G 2010 *Anal. Chem.* **82** 6652–9

[32] Boonrungsiman S, Fearn S, Gentleman E, Spillane L, Carzaniga R, McComb D W, Stevens M M and Porter A E 2013 *Nanoscale* **5** 7544–51

[33] Jones E A, Lockyer N P and Vickerman J C 2008 *Anal. Chem.* **80** 2125–32

[34] Ryan J L 1996 The atmospheric deterioration of glass: studies of decay mechanisms and conservation techniques *PhD Thesis* University of London

[35] Fearn S, McPhail D S and Oakley V 2005 *Phys. Chem. Glasses* **46** 505–11

[36] Lodding A, Odelius H, Clark D E and Werme L O 1985 *Mikrochim. Acta* 11 (Suppl.) p 145

[37] Wicks G G 1991 Nuclear waste glasses: corrosion behaviour and field tests *Corrosion of Glass, Ceramics, and Superconductors* ed D E Clark and B K Zoitos (Park Ridge, NJ: Noyes)

[38] Jain V and Pan Y M 2000 Glass melt chemistry and product qualification *Report for the Centre for Nuclear Waste Regulatory Analyses (CNWRA) San Antonio, TX, USA. Nuclear Regulatory Commission Contract NRC-02-97-009*

[39] Lodding A and Van Iseghem P 2001 *J. Nucl. Mater.* **298** 197–202

[40] Chave T, Frugier P, Ayral A and Gin S 2007 *J. Nucl. Mater.* **362** 466–73

[41] Ahmad N E, Fearn S, Jones J R and Lee W E 2014 *Procedia Mater. Sci.* **7** 230–36

[42] Fearn S 2000 A SIMS based bevel-image technique for the analysis of semiconductor materials *PhD Thesis* University of London

[43] Kilner J A, Steele B C H and Ilkov L 1984 *Solid State Ionics* **12** 89–97

[44] Carter S, Selcuk A, Chater R J, Kajda J, Kilner J A and Steele B C H 1992 *Solid State Ionics* **53–6** 597–605

[45] DeSouza R A, Zehnpfenning J, Martin M and Maier J 2005 *Solid State Ionics* **176** 1465–71

[46] Crank J 1975 *The Mathematics of Diffusion* 2nd edn (Oxford: Oxford University Press)

[47] MATLAB R13 (Natick, MA: Mathworks)

[48] DeSouza R and Martin M 2009 *MRS Bull.* **34** 907–14

[49] Schulz O, Flege S and Martin M 2003 *The ECS Proceedings Series: SOFC VIII* eds S C Singhal and M Doykia p 304

[50] Pergolesi D, Fabbri E, Cook S N, Roddatis V, Traversa E and Kilner J A 2012 *ACS Nano* **6**(12) 10524–34

[51] Saranya A M, Pla D, Morata A, Cavallaro A, Canales-Vazquez J, Kilner J A, Burriel M and Tarancon A 2015 *Adv. Energy Mater.* **5** DOI: 10.1002/aenm.201500377

[52] Wilson R G, Stevie F A and Magee C W 1989 *Secondary Ion Mass Spectrometry: a Practical Handbook for Depth Profiling and Bulk Impurity Analysis* (New York: Wiley)

[53] Gui D, Shao J J, Hao M, Xing Z X, Lee H S, Shen Y Q, Li X M and Cha L Z 2014 *Surf. Int. Anal.* **46** 307–11

[54] Schnieders A 2010 Full wafer defect analysis with time-of-flight secondary ion mass spectrometry *IEEE/SEMI Advanced Semiconductor Manufacturing Conf. (ASMC) (San Francisco, CA, 11–13 July 2010)* pp 158–61

[55] Gorbenko V *et al* 2015 *Phys. Status Solidi RRL* **9** 202–5

[56] www.intel.com/content/www/us/en/silicon-innovations/standards-22nm-3d-tri-gate-transistors-presentation.html

[57] Treat N D, Westacott P and Stingelin N 2015 *Annu. Rev. Mater. Res.* **45** 15.1–32

[58] Westacott P *et al* 2013 *Energy Environ. Sci.* **6** 2756–64

[59] Yu G, Gao J, Hummelen J C, Wudl F and Heeger A J 1995 *Science* **270** 1789–91

[60] Treat N D, Brady M A, Smith G, Toney M F, Kramer E J, Hawker C J and Chabinye M L 2011 *Adv. Energy Mater.* **1** 82–9

[61] Pfannmöller M *et al* 2011 *Nano Lett.* **11** 3099–107

[62] Westacott P 2013 Emerging semiconductor technologies—focusing on photovoltaics *PhD Thesis* Imperial College London

[63] Smentkowski V S, Zorn G, Misner A, Parthasarathy G, Couture A, Tallarek E and Hagenhoff B 2013 *J. Vac. Sci. Technol.* A **31** 30601–6

[64] Bailey J, Havelund R, Shard A G, Gilmore I S, Alexander M R, Sharp J S and Scurr D J 2015 *ACS Appl. Mater. Interfaces* **7** 2654–9

[65] Busby Y, Crespo-Monteiro N, Girleanu M, Brinkman M, Ersen O and Pireaux J-J 2015 *Organic Electronics* **16** 40–5

[66] Thompson R J, Fearn S, Tan K-J, Cramer H-G, Kloc C L, Curson N J and Mitrofanov O 2013 *Phys. Chem. Chem. Phys.* **15** 5202–7

An Introduction to Time-of-Flight Secondary Ion
Mass Spectrometry (ToF-SIMS) and its Application
to Materials Science

Sarah Fearn

Chapter 6

Summary

Dual-beam ToF-SIMS is an extremely versatile analytical tool that can be successfully used for the characterisation of an ever-widening range of materials. Its unique combination of depth profiling, ion imaging and static mode analysis, with very high sensitivity, can generate a huge amount of information with regard to elemental, molecular and chemical composition not only on two dimensions, but also three. This introduction to dual-beam ToF-SIMS has presented a brief summary of the SIMS technique, its basic principles, and recent instrumental developments along with some applications to materials science. It is by no means exhaustive.

The decades of instrumental development has also paid off, with the introduction of large cluster ion sources; SIMS can now be usefully applied in areas of material research that would have been unthinkable even five years ago. As the understanding of the interactions between the organic-based materials and argon cluster ion beams grows, it will not be long until the depth profiling of multi layer polymer samples is commonplace, and quantifiable data readily attainable.

The adoption of SIMS as an analytical technique in the area of the biological sciences is also extremely exciting, as it is able to exploit the strength of the SIMS to provide spatially resolved chemistry. In this domain, however, sample preparation can still be difficult for delicate samples, and it may be that more *in situ* preparation techniques need to be developed. The possibility of being able to quantify the data in this arena would also be a great breakthrough. Hopefully as the technique is exposed to more and more biological questions, a realisation of suitable standards may be developed.

Across all fields of materials research, the challenge in utilising ToF-SIMS is to exploit the versatility that its unique properties provide, but to also couple this with a deep understanding of its limitations, to ensure it is applied usefully to obtain accurate and precise information.

www.ingramcontent.com/pod-product-compliance
Lightning Source LLC
Chambersburg PA
CBHW041451210326
41599CB00004B/206